W0081280

"This new book by Pereira and Viola shows remarkable breadth and depth. It reveals that the Amazon forest is approaching a deforestation tipping point, with high stakes for not just global climate change, but also biodiversity conservation. By focusing on the politics behind the outcomes – in four Amazonian countries and internationally – the book shows both how the region reached this threshold and also how it could still pull back from the brink. A tour de force!"

Kathryn Hochstetler, *Professor and Head of Department of International Development, London School of Economics and Political Science*

"The Amazon is the Earth's most crucial ecosystem, and it is in grave crisis. With the looming prospect of climate change and habitat destruction combining to trigger a disastrous ecological "tipping point", Pereira and Viola offer a detailed and incisive analysis of how regional governments, parties and corporations are pushing the Amazon to the brink of catastrophe. This book reminds us that the Anthropocene is not merely a dystopian playground for social theory but an urgent crisis of the biosphere and failing systems of greed, extraction, and governance. It should be read by every environmentalist, scholar and politician with a conscience."

Anthony Burke, *Professor of Environmental Politics and International Relations, University of New South Wales, Canberra*

"Pereira and Viola provide an authoritative overview of the political processes that have shaped the environmental governance of the Amazon basin in the last two decades. While the book provides a useful summary of Brazilian policies and politics, its main contribution lies in a systematic analysis of Peru, Bolivia and Colombia, countries that are often overlooked. As such, the book provides a truly regional overview of the complex set of issues affecting the future of the rainforest and the global climate."

Raoni Rajão, *Associate Professor in Environmental Management, Federal University of Minas Gerais (UFMG)*

"Humans can live with one lung, yet they are born with two. Nature has been generous with us: by means of redundancy, it chose safety over efficiency. Yet the world has one lung, and it is losing it – or rather, we humans are wrecking it. Through extensive fieldwork in four South American countries, Pereira and Viola scrutinized biodiversity governance in the Amazon to measure the extent of the damage and suggest policy options. Their work is as scholarly solid as it is politically compelling."

Andrés Malamud, *Principal Researcher, Institute of Social Sciences, University of Lisbon*

Climate Change and Biodiversity Governance in the Amazon

This book provides an analysis of the recent governance of the Amazon in Brazil, Peru, Bolivia and Colombia, with a particular focus on deforestation processes, demonstrating that current policies and political and socioeconomic dynamics in the four countries are risking the forest's resilience.

The authors examine and compare Amazonian politics and policies under different administrations, concentrating on the main actors, policies and dynamics that have affected the region, as well as on the institutional and political environment in which deforestation processes were embedded in different periods. Essentially, the book makes an analytical contribution towards a better understanding of the political, economic and social challenges confronting conservation policy in the Amazonian countries.

Climate Change and Biodiversity Governance in the Amazon: At the Edge of Ecological Collapse? is essential reading for students and researchers in the fields of environmental studies and sustainability, Latin American studies, political science and international relations, as well as for policymakers and practitioners working in conservation and development.

Joana Castro Pereira is a researcher at the Portuguese Institute of International Relations (IPRI), Universidade Nova de Lisboa, Portugal.

Eduardo Viola is a full professor at the Institute of International Relations (IREL), Universidade de Brasília, and a senior research fellow at the Institute of Advanced Studies (IEA), Universidade de São Paulo, Brazil.

Routledge Advances in International Relations and Global Politics

United Nations Financial Sanctions
Edited by Sachiko Yoshimura

Operational Code Analysis and Foreign Policy Roles
Crossing Simon's Bridge
Edited by Mark Schafer and Stephen G. Walker

Social Media Impacts on Conflict and Democracy
The Techtonic Shift
Edited by Lisa Schirch

Changing Arms Control Norms in International Society
Kenki Adachi

The Frontiers of Public Diplomacy
Hegemony, Morality and Power in the International Sphere
Colin R. Alexander

Security and Safety in the Era of Global Risks
Edited by Radomir Compel and Rosalie Arcala Hall

Laws of Politics
Their Operations in Democracies and Dictatorships
Alfred G. Cuzán

Climate Change and Biodiversity Governance in the Amazon
At the Edge of Ecological Collapse?
Joana Castro Pereira and Eduardo Viola

For information about the series: https://www.routledge.com/Routledge-Advances-in-International-Relations-and-Global-Politics/book-series/IRGP

Climate Change and Biodiversity Governance in the Amazon

At the Edge of Ecological Collapse?

Joana Castro Pereira and Eduardo Viola

NEW YORK AND LONDON

First published 2022
by Routledge
605 Third Avenue, New York, NY 10158

and by Routledge
2 Park Square, Milton Park, Abingdon, Oxon, OX14 4RN

Routledge is an imprint of the Taylor & Francis Group, an informa business.

Library of Congress Cataloging-in-Publication Data
Names: Pereira, Joana Castro, author. | Viola, Eduardo J., author.
Title: Climate change and biodiversity governance in the Amazon : at the edge of ecological collapse? / Joana Castro Pereira, Eduardo Viola.
Description: New York, NY : Routledge, 2021. | Series: Routledge advances in international relations and global politics | Includes bibliographical references and index. | Identifiers: LCCN 2021009431 (print) | LCCN 2021009432 (ebook) | ISBN 9780367275549 (hardback) | ISBN 9781032058801 (paperback) | ISBN 9780429296581 (ebook) | ISBN 9781000428278 (adobe pdf) | ISBN 9781000428292 (epub)
Subjects: LCSH: Environmental policy--Amazon River Region. | Climatic changes--Amazon River Region. | Deforestation--Amazon River Region. | Biodiversity--Amazon River Region. | Amazon River Region--Environmental conditions. Classification: LCC GE190.A43 P47 2021 (print) | LCC GE190.A43 (ebook) | DDC 363.738/7456109811--dc23
LC record available at https://lccn.loc.gov/2021009431
LC ebook record available at https://lccn.loc.gov/2021009432

ISBN: 978-0-367-27554-9 (hbk)
ISBN: 978-1-032-05880-1 (pbk)
ISBN: 978-0-429-29658-1 (ebk)

Typeset in Times New Roman
by MPS Limited, Dehradun

Contents

Acknowledgments

The authors would like to thank Mikaela Weisse, Anika Berger and James MacCarthy from the Global Forest Watch for their valuable help in providing forest loss data for the Amazonian countries addressed in this book, and kind availability to promptly respond to our queries; the interviewees, who generously gave of their time to participate in this research, sharing not only their views, knowledge and experience, but also documents and data, and without whose contribution this book would have not been possible; Miguel Rodrigues Freitas for his insightful and constructive comments on earlier versions of the various chapters; and José Félix Pinto-Bazurco, Jenny Gruenberger and Carlos Castaño-Uribe for their attentive reading of the chapters on Peru, Bolivia and Colombia, respectively. Any errors remain our own.

Funding for this work was provided by the Portuguese Foundation for Science and Technology (FCT) in the framework of the project CEECIND/00065/2017, and by the Brazilian National Council for Scientific and Technological Development (CNPq).

List of Abbreviations/Acronyms

ABS	Access and Benefit-Sharing
ABT (Spanish acronym)	Forest and Land Authority (Bolivia)
AIDESEP (Spanish acronym)	Inter-Ethnic Association for the Development of the Peruvian Jungle
BPBES (Portuguese acronym)	Brazilian Platform on Biodiversity and Ecosystem Services
CAN (Spanish acronym)	National Agrarian Commission (Bolivia)
CAR (Spanish acronym)	Regional Autonomous Corporations (Colombia)
CBD	Convention on Biological Diversity
CBDRRC	Common but Differentiated Responsibilities and Respective Capabilities
CCP	Company-Community Partnerships
CEPLAN (Spanish abbreviation)	National Center for Strategic Planning (Peru)
CIBOD (Spanish acronym)	Confederation of Indigenous Peasants of Bolivia
CNA (Portuguese abbreviation)	Brazilian Confederation of Agriculture and Livestock
CONAMA (Portuguese acronym)	National Environmental Council (Brazil)

CONAMAQ (Spanish acronym)	National Council of Allyus and Markas of the Qullasuyu (Bolivia)
CONFIEP (Spanish abbreviation)	National Confederation of Private Business Institutions (Peru)
COP	Conference of the Parties
DGFFS (Spanish acronym)	General Directorate for Forestry and Wildlife (Peru)
DPP	Democratic Prosperity Policy (Colombia)
DSP	Democratic Security Policy (Colombia)
DSPLEE	Defense and Security Policy for Legality, Entrepreneurship and Equity (Colombia)
ELN (Spanish acronym)	National Liberation Army (Colombia)
ENGOs	Environmental Non-Governmental Organizations
FARC (Spanish acronym)	Revolutionary Armed Forces of Colombia
FEDEMIN (Spanish abbreviation)	Federation of Miners of Madre de Dios (Peru)
FES	Socio-Economic Function (Bolivia)
FPA (Portuguese acronym)	Agribusiness Parliamentary Front (Brazil)
GDAEI (Spanish acronym)	General Directorate of Agrarian Environmental Issues (Peru)
GFW	Global Forest Watch
GHG	Greenhouse Gas
GSPC	Global Strategy for Plant Conservation
GtC	Gigatons of Carbon
GTRC (Spanish acronym)	Territorial Management with Shared Responsibility (Bolivia)

IBAMA (Portuguese abbreviation)	Brazilian Institute of Environment and Renewable Natural Resources
ICMBio (Portuguese abbreviation)	Chico Mendes Institute for Biodiversity Conservation (Brazil)
IDEAM (Spanish abbreviation)	Institute of Hydrology, Meteorology and Environmental Studies (Colombia)
INDEPA (Spanish acronym)	National Institute for the Development of Andean, Amazonian and Afro-Peruvian Peoples
INRA (Spanish acronym)	National Service of Agrarian Reform (Bolivia)
IPBES	Intergovernmental Science-Policy Platform on Biodiversity and Ecosystem Services
IPCC	Intergovernmental Panel on Climate Change
LULUCF	Land Use, Land-Use Change and Forestry
MAAP	Monitoring of the Andean Amazon Project
MAS (Spanish acronym)	Movement Towards Socialism (Bolivia)
MEF (Spanish acronym)	Ministry of Economy and Finance (Peru)
MINAGRI (Spanish abbreviation)	Ministry of Agriculture (Peru)
MINAM (Spanish abbreviation)	Ministry of Environment (Peru)
MINEM (Spanish abbreviation)	Ministry of Energy and Mines (Peru)
MMA (Portuguese acronym)	Ministry of Environment (Brazil)
MMAyA (Spanish acronym)	Ministry of Environment and Water (Bolivia)
NDC	Nationally Determined Contribution

NGOs	Non-Governmental Organizations
OECD	Organization for Economic Co-operation and Development
OEFA (Spanish acronym)	Environmental Assessment and Enforcement Agency (Peru)
OSINFOR (Spanish abbreviation)	Agency for the Supervision of Forest Resources and Wildlife (Peru)
OTCA (Portuguese/Spanish acronym)	Amazon Cooperation Treaty Organization
PCM (Spanish acronym)	Presidency of the Council of Ministers (Peru)
PNCBMCC (Spanish acronym)	National Forest Conservation Program for the Mitigation of Climate Change (Peru)
PPCDam (Portuguese acronym)	Action Plan for the Prevention and Control of Deforestation in the Legal Amazon (Brazil)
PT (Portuguese acronym)	Workers' Party (Brazil)
RAISG (Spanish acronym)	Amazon Geo-Referenced Socio-Environmental Information Network
REDD	Reducing Emissions from Deforestation and Forest Degradation
REDD+	Reducing Emissions from Deforestation and Forest Degradation and the role of conservation, sustainable management of forests and enhancement of forest carbon stocks in developing countries
SENACE (Spanish abbreviation)	National Service of Environmental Certification for Sustainable Investments (Peru)

SERFOR (Spanish abbreviation)	National Forest and Wildlife Service (Peru)
SERNANP (Spanish acronym)	National Service of Natural Protected Areas (Peru)
SERNAP (Spanish acronym)	National Service of Protected Areas (Bolivia)
SMByC (Spanish acronym)	Forest and Carbon Monitoring System (Colombia)
SPDA (Spanish acronym)	Peruvian Environmental Law Society
SRB (Portuguese acronym)	Brazilian Rural Society
TIPNIS (Spanish acronym)	Isiboro Securé National Park and Indigenous Territory (Bolivia)
TPA	Trade Promotion Agreement
TPGRFA	International Treaty on Plant Genetic Resources for Food and Agriculture
TRIPS	Agreement on Trade-Related Aspects of Intellectual Property Rights
UNASUR (Spanish abbreviation)	Union of South American Nations
UNFCCC	United Nations Framework Convention on Climate Change
ZIDRES (Spanish acronym)	Agro-Industrial Special Zones (Colombia)
ZRC (Spanish acronym)	Peasant Reserve Areas (Colombia)

1 Introduction*,[1]

Extending through Brazil, Peru, Bolivia, Colombia, Venezuela, Guyana, Suriname, Ecuador and French Guyana, the Amazon rainforest is the greatest continuous tropical forest in the world, covering over 8 million km^2 (RAISG, 2020).[2] It stores 150–200 gigatons of carbon (GtC), that is, roughly half of all tropic forest carbon (Brienen et al., 2015), and possibly hosts a quarter of the planet's terrestrial species (The World Bank, 2019). The Amazon is a key element of the Earth system. The dieback of the forest could have catastrophic consequences.

Several studies published in recent years suggest that the resilience of the Amazon is deteriorating, potentially leading to abrupt and irreversible environmental damage during this century (see, for instance, Baccini et al., 2017; Brienen et al., 2015; Nobre et al., 2016; Staal et al., 2020; Zemp, Schleussner, Barbosa, & Rammig, 2017). If this is the case, not only would local populations be severely affected; massive amounts of carbon would be released into the atmosphere, further exacerbating global warming, and biodiversity would be lost, as many terrestrial species would lose their natural habitats, thus precipitating an environmental catastrophe. In addition, the Amazon's hydrological engine, which plays a critical role in regulating the planet's climate, would likely be modified. Of further concern is the fact that the current hydrological regime sustains agricultural production in Brazil, which is the world's second largest food exporter. In sum, the deterioration of the Amazon's resilience may trigger a cascading effect that would completely alter the Earth's water cycle and climate, and threaten global food security (Cardil et al., 2020). Unique species of plants and animals – which, in addition to their intrinsic value as living organisms, support healthy ecosystems that humans rely on – could disappear. However, neither the social science community that is studying the Amazon nor the political

*This introduction draws partly on Pereira and Viola (2020).

debates over the region have yet acknowledged the real prospect of the dieback of the forest (Pereira & Viola, 2018, 2019). The "Amazon tipping point" warrants urgent political attention (Amigo, 2020).

1.1 The Amazon Tipping Point: Internal and External Threats to the Forest's Resilience

A tipping point is the "point or threshold at which small quantitative changes in the system trigger a non-linear change process that is driven by system-internal feedback mechanisms and inevitably leads to a qualitatively different state of the system, which is often irreversible" (Milkoreit et al., 2018, p. 9). Two tipping points have been associated with the collapse of large parts of the Amazon. In the absence of other contributing factors, a global temperature increase of 4 °C could cause the forest to enter a catastrophic feedback loop, transforming forest ecosystems into impoverished savannas. Considering potential negative synergies between deforestation, climate change and widespread use of fire, estimates indicate a threshold for the dieback of the forest at 20% to 25% deforestation (Lovejoy & Nobre, 2018). Alarmingly, the region has warmed by approximately 1 °C over the last six decades, and deforestation is close to reaching 20% of the forested area (Nobre et al., 2016). According to a recent study by Staal et al. (2020), in approximately 40% of the Amazon, the rainfall is currently at a level where the forest could shift to a savanna. In parts of the Amazon, the signs of savannization are already visible; while some believe that the process is still reversible if global warming is mitigated and the forest is restored, others are more pessimistic (see Chiaretti, 2021).

Global temperatures have been rising at a rate that challenges previous climate models. The decade 2011–2020 was the warmest on record. The warmest six years have all occurred since 2015. In 2020, the global average temperature was nearly 1.2 (± 0.1) °C above pre-industrial levels (WMO, 2021). Since 2013, land temperatures have persistently registered an annual average anomaly above 1 °C, reaching +1.59 °C in 2020 (NOAA, n. d.). The Amazon has been severely hit by rising global temperatures; drought–fire interactions are causing tree mortality and biomass loss in the region (Brando et al., 2014). Climate change is reducing the capacity of disturbed forests to recover and recapture lost carbon (Exbrayat, Liu, & Williams, 2017). As the planet warms, the region will certainly experience more frequent and severe droughts and fires that may further jeopardize the forest's resilience (De Faria et al., 2017; Duffy, Brando, Asner, & Field, 2015; Staal et al., 2020).

Anthropogenic disturbances in the Amazon such as forest conversion to agricultural use, logging and other resource extraction, deliberate fires, the hunting and wildlife trade, and the spread of introduced species and pathogens are acting in synergy with the effects of climate change, potentially leading to outcomes that cannot be predicted (Malhi, Gardner, Goldsmith, Silman, & Zelazowski, 2014).

Deforestation has increased in most of the Amazon in recent years; environmentally-damaging activities have intensified in many of the region's countries. Additionally, every year over the last two decades, an area almost the size of Uruguay (169,000 km^2) has been affected by burning and forest fires. Bolivia and Brazil are the countries which have suffered the greatest impact of land burning and forest fires, with 27% and 17% respectively (RAISG, 2020).

Preventing the Amazon from tipping into a savanna will require not only good governance at national level, but also the promotion of a new, sustainable development paradigm in the region as well as progressive environmental diplomacy by the Amazonian countries in the international climate and biodiversity regimes, as national and regional action is insufficient to protect the forest. It is absolutely essential that the countries of the Amazon basin work within the United Nations Framework Convention on Climate Change (UNFCCC) towards ensuring that the global temperature does not reach a level that causes the collapse of large parts of the Amazon. It is also critical that those countries become active, vocal players within the Convention on Biological Diversity (CBD), pressuring for more robust strategies and mechanisms to conserve and make a better, socially-inclusive use of biodiversity on a global scale. The Amazonian countries in both regimes should work together to build solid coalitions and partnerships to manage and preserve the forest as well as attracting the best investments to develop new initiatives in the region that could promote the conservation and sustainable use of the components of biological diversity, and improve economic and living opportunities for local communities (see Pereira & Viola, 2020). Finally, those countries should also influence international negotiations towards a close articulation between the UNFCCC and the CBD, as the lack of integration between the two regimes is a major limitation. Climate change and biodiversity loss are deeply interconnected – climate change is one of the main drivers of biodiversity loss, and the loss of primary forests exacerbates climate change. The urgency of such active and progressive diplomacy is reinforced by the insufficiency of global efforts to fight global warming and species loss.

Recently, a number of large greenhouse gas (GHG) emitters announced net-zero pledges: China for 2060 and the United States, the European Union, Japan, South Korea, Canada and South Africa for 2050. A total of 127 countries accounting for over 60% of emissions are considering or have adopted net-zero emission targets. These optimistic targets would only limit global warming to approximately 2.1 °C by 2100, still failing to meet the "well below 2 °C" and the 1.5 °C aspirational temperature goals established in the Paris Climate Agreement; current pledges under the agreement and policies would potentially lead to a global warming increase of 2.6 °C and 2.9 °C, respectively, by the end of the century. Stronger 2030 targets are urgently needed to close the emissions gap (CAT, 2020). According to the 2018 special report by the Intergovernmental Panel on Climate Change (IPCC, 2018), the global carbon budget for keeping global warming at 1.5 °C could be depleted in a few years. Nevertheless, the IPCC's latest conclusions have been widely criticized for being too conservative. In fact, there is a growing consensus among climate scientists that the threshold between incremental and dangerous climate change has already been crossed; global temperatures could exceed 1.5 °C by the early 2030s (Henley & King, 2017). This is alarming for the Amazon, as in a 2 °C global warming scenario, more than one-third of the region's wildlife would be at risk of extinction (Warren et al., 2018).

Within the CBD, the strategic plan for biodiversity for the 2011–2020 period failed; none of the 20 targets to stop the destruction of nature and species extinctions were fully met at global level (Secretariat of the Convention on Biological Diversity, 2020), which points to the urgent need to rethink and transform the global governance of biodiversity. The sixth mass extinction is underway. The basic fact that biodiversity is the foundation of humanity's life support system remains dramatically underrecognized in political discussions and within the policy domain (Lenton, 2019).

Both the UNFCCC and the CBD are failing the Amazon and the planet, but if the Amazonian countries want to promote an ambitious global response to the climate and biodiversity crises that is capable of protecting the Amazon and all other ecosystems on Earth, they themselves will need first to rethink their relationship with the forest.

1.2 Aim and Outline of the Book, Methodology and Other Relevant Considerations

This book provides an analysis of the recent governance of the Amazon in Brazil (2005–2019), Peru (2006–2019), Bolivia (2006–2019) and Colombia (2002–2019), with a particular focus on the political determinants of deforestation in the four countries. It examines and compares Amazonian politics and policies under different administrations, concentrating on the main actors, policies and dynamics that have affected the region, as well as on the institutional and political environment in which deforestation processes were embedded in different periods. By investigating the challenge of sustainable development in four Amazonian countries, which together account for approximately 90% of the forest, we argue that the "savannization hypothesis" is potentially closer to reality than most political and academic debates in the social sciences assume, and must be seriously considered. Our analysis demonstrates that current Amazonian policies are critically inadequate for safeguarding the forest's resilience.

Nevertheless, the analysis was not a simple one for two reasons. First, since our research focuses on politics and policies, we are particularly interested in the *human* causes of (primary)[3] forest loss in the Amazon, that is, *deforestation*. However, we found widespread undifferentiated use of the terms "deforestation" and "forest loss" – note that the latter includes *both human and natural* disturbances of woody vegetation. The Brazilian PRODES program is most likely the most accurate source of deforestation data for the region,[4] essentially because their technicians conduct a manual validation of the satellite images used, which allows for a more confident selection of primary forest loss areas by *anthropogenic action*. However, PRODES only produces data for the Brazilian Amazon. When we began our study, we were using, in addition to PRODES, deforestation data produced by the Amazon Geo-Referenced Socio-Environmental Information Network (RAISG, 2016), since it was the only deforestation data source available for all Amazonian countries, following the same methodology. This would allow us to more rigorously compare the different countries' case studies. Yet, when we started interviewing experts for the Colombian case, doubts were raised about the deforestation trends presented by RAISG. For example, for the 2005–2010 period in Colombia, data by RAISG (2016) indicated an increase of almost 80% in Amazonian deforestation levels compared to the previous five-year period. Despite high uncertainty among the interviewees for the Colombian case on deforestation trends over different periods, several of them found such a massive increase odd. Similar doubts emerged for Peru and Bolivia. Confusion was further exacerbated by the interchangeable use of the concepts of "deforestation" and "forest loss". Meanwhile, RAISG published new deforestation data in December 2020,[5] when this book was being finished; the new data, produced according to a different methodology, indicates significantly different trends for some countries in relation to the data published in 2016. In light of this uncertainty, we opted to combine different data from various sources in our analysis and tried to find similar, general trends among them. The Brazilian case is the only one for which clear, relatively congruent trends emerged among the different sources consulted. For the other countries' case studies, it is more difficult to associate a particular policy or event to the figures available, as there are significant differences among the various sources (see Appendix A). All data sources consulted other than the PRODES program do not perform, as far as we were able to understand, any kind of manual image photointerpretation to identify deforested areas, which precludes a more rigorous distinction between human and natural causes of forest loss. More information on the probable factors that have led to differences between data from the various sources consulted can be found in Appendix A.

Second, deforestation processes in the Peruvian and Colombian Amazon are still not fully understood, which complicated our political analysis of both cases. More details on our methodological options for each country are provided in the respective chapters. It seems clear, nonetheless, from the data presented in Appendix A, alongside the empirical evidence collected in this book, that in recent years deforestation has increased in most of the Amazon. This supports our argument for the possibility of crossing a tipping point during this century, and the need to deepen our knowledge of the political and socioeconomic factors fueling the Amazon's destruction.

As for structure, in addition to this introduction and the conclusion, the book is divided into four chapters. Chapter 2 addresses the Brazilian case. It analyzes the political, economic and social factors underlying the trajectory of Amazonian deforestation in the country in the period 2005 to 2019. It also examines the Brazilian diplomatic performance at the UNFCCC and the CBD. As mentioned in the previous subsection, protecting the Amazon depends also on collective action on a global scale. Moreover, Brazil is a relevant country in global governance, a central actor in the planetary carbon cycle and a critical player for the stability of the Earth system, being the world's seventh largest emitter and holding the planet's most important forest carbon stock, largest biodiversity and reserve of agricultural land (Viola & Franchini, 2018). The country accounts for over 60% of the Amazon and has significant scientific, institutional and financial capacity vis-à-vis its neighboring countries. Consequently, any strategy for protecting the Amazon on an international level will require Brazilian leadership (Pereira & Viola, 2020). Understanding the country's stance in the climate and biodiversity regimes is thus fundamental. Chapters 3, 4 and 5 focus respectively on the governance of the Amazon in Peru, Bolivia and Colombia. In light of the uncertainties regarding deforestation trends for those countries, the analyses presented in these chapters are less concise than the one for Brazil, including a discussion of a wider range of factors and events with potential Amazonian implications, both negative and positive. For this reason, and due to the limit on space, we did not include an analysis of the Peruvian, Bolivian and Colombian positions within the UNFCCC and the CBD. The book concludes in Chapter 6 with a summary of the main dynamics that have driven deforestation in the four countries.

The study is mainly based on 82 semi-structured interviewees with an average duration of 1 hour 30 minutes, conducted between May and October 2020, with key public officials from different ministries and governmental agencies (39), academics with diverse backgrounds (23) and environmental non-governmental organization (ENGO) members working in the Amazon (20). All interviewees were anonymized to protect their identities. We omitted the day when each interview took place in the list of interviews referenced in each chapter, as that information could identify some of the participants. Of the 82 interviews, only two were conducted for the Brazilian case; these focused on the country's position on the CBD and consisted mostly of advisory conversations, as there is a significant amount of research published in Portuguese on the topic. Moreover, given our accumulated knowledge of Brazilian Amazonian politics and policies and the country's participation at the UNFCCC's negotiations, obtained over more than three decades of research on both themes, as well as the availability of solid deforestation figures and relatively congruent deforestation trends among different sources for the Brazilian Amazon, we did not feel the need to conduct more interviews for the Brazilian case. The purpose of the remaining 80 interviews was to collect insights and qualitative data on the trends and political drivers of, and the social and economic dynamics affecting, Amazonian deforestation in Peru, Bolivia and Colombia, and how the politics and policies of different administrations impacted on the region as well as the main continuities and changes between those administrations. The interviews were triangulated with and complemented by each other, in addition to governmental publications and data, news reports published in national and international newspapers, national and international scientific literature, and studies conducted by scientific agencies, consortiums and international organizations.

1.3 Our Contribution to the Literature

Despite the forest's global importance, the study of Amazonian governance is virtually absent from the literature written in English in the fields of Political Science and International Relations. There is a significant amount of research on the local governance of the Amazon in Brazil written in Portuguese, and a more limited amount of work on the topic regarding the other Amazonian countries, written mostly in Spanish. A considerable number of articles have been published over the past decade on Bolivia, analyzing the profound contradictions between governmental ecologist rhetoric and legislative initiatives recognizing nature's rights on the one hand, and the deepening of extractivist policies on the other hand, as well as the dynamics of agro-extractivism in the countryside. However, those articles focus neither on the governance of the Amazon nor on explaining deforestation processes. In addition, a few reports on forest governance in the region by national and international research organizations have been published; nevertheless, these do not explain policy developments. With the exception of a few countries, namely Brazil, Mexico, Chile and Costa Rica, research on Latin American environmental policy is scarce. This is a substantial gap, considering the region's exceptional natural wealth; eight of the 20 most biodiverse countries in the world are located in Latin America (Butler, 2016). This book sheds light on the actors involved in environmental policy, on their interests and power, examining the interactions between them to explain policy outcomes and deforestation processes in the Amazonian countries which account for the largest proportion of the forest. It looks at national government executive, legislative and bureaucratic institutions as well as the material (e.g., individual firms, business associations, specific industries) and ideational (e.g., ENGOs, political parties, public opinion) forces that operate either to advance or obstruct the national environmental agenda; it also notes the influence of international actors and factors on domestic policies. In addition, since the book analyzes the evolution of national politics and policies, state institutions and actors over a certain period of time for each of the countries addressed, thus covering several political administrations and identifying the main patterns of similarity and difference, it also makes a contribution to the field of environmental comparative politics. Regarding the Brazilian case in particular, this book also brings research written in Portuguese to a wider public, on the country's position at the CBD, the domestic dynamics that explain such a stance and the impact of developments on the biodiversity regime in the country. Ultimately, the book makes an analytical contribution towards a better understanding of the political, economic and social challenges confronting conservation policy in the Amazonian countries.

Moreover, by demonstrating that the Amazon is risking abrupt and irreversible environmental damage during this century, this book defies dominant analyses which assume progressive and linear processes of environmental degradation (see Pereira & Viola, 2018). Scholars in the social sciences and decisionmakers are prevented from recognizing the growing possibility of a global ecological catastrophe by ignoring (a) the fact that anthropogenic interference with the Earth system may have already crossed the boundary below which the danger of destabilization of the system is likely to remain low; and (b) the dominant role humanity currently plays in driving ecological change (Dryzek & Pickering, 2019). As will be seen, this risk is the result of human, political decisions; consequently, mitigating it is fundamentally a matter of implementing new policies and reorienting human activities (Burke, Fishel, Mitchell, Dalby, & Levine, 2016). Our comprehensive analysis of the broad dynamics affecting the ecological integrity of the Amazon concurs with recent calls for recognizing and analyzing "the inherent intertwinement between world politics and the Earth system (…)[;] the fundamental embeddedness of the 'human' in the 'non-human part of the universe', and how every manifestation of world politics is necessarily entangled with non-human processes and species" (Pereira & Saramago, 2020, p. 4). Here, human/non-human nature relations are empirically conceived as an entangled whole, thus rejecting the prevailing dualism between humans and nature.

Notes

1 An area considering the biome, the hydrographic basin and the administrative regions of Brazil and Ecuador (RAISG, 2020).

2 Defined as "a biomass loss of at least 25% of the total biomass in 2070–2100 in comparison to 1970–2000" (Rammig et al., 2010, p. 698).

3 Primary forests "store about 30–50% more carbon than degraded forests (…). Their carbon stocks are more stable and resilient than those of degraded forests or plantations because their [native] biodiversity and ecosystem resilience make them more resistance to external pressures. (…) Primary forests are irreplaceable. (…). [T]hey protect the most carbon and biodiversity, produce the cleanest freshwater, regulate water flows, have local cooling effects and prevent erosion" (Kormos, Mackey, Mittermeier, & Young, 2020).

4 The PRODES program defines deforestation as the "conversion by suppression of areas of primary forest physiognomy by anthropogenic actions" [authors' translation from Portuguese] (INPE, 2019, p. 4). This definition is in line with the aim and scope of our analysis.

5 See https://www3.socioambiental.org/geo/RAISGMapaOnline/

References

Amigo, I. (2020). When Will the Amazon Hit a Tipping Point? *Nature*, *578*, 505–507.

Baccini, A., Walker, W., Carvalho, L., Farina, M., Sulla-Menashe, D., & Houghton, R.A. (2017). Tropical Forests Are a Net Carbon Source Based on Aboveground Measurements of Gain and Loss. *Science*, *358*, 6360. doi:10.1126/science.aam5962

Brando, P.M., Balch, J.K., Nepstad, D.C., Morton, D.C., Putz, F.E., Coe, M.T.,... Soares-Filho, B.S. (2014). Abrupt Increases in Amazonian Tree Mortality Due to Drought–Fire Interactions. *Proceedings of the National Academy of Sciences*, *111*(17), 6347–6352. doi:10.1073/pnas.1305499111

Brienen, R.J., Phillips, O.L., Feldpausch, T.R., Gloor, E., Baker, T.R., Lloyd, J.,... Martinez, R.V. (2015). Long-Term Decline of the Amazon Carbon Sink. *Nature*, *519*, 344–348. doi:10.1038/nature14283

Burke, A., Fishel, S., Mitchell, A., Dalby, S., & Levine, D.J. (2016). Planet Politics: A Manifesto from the End of IR. *Millennium: Journal of International Studies*, *44*(3), 499–523. doi:10.1177/0305829816636674

Butler, R.A. (2016, May 21). The Top 10 Most Biodiverse Countries. *Mongabay Latam*. Retrieved from https://news.mongabay.com/2016/05/top-10-biodiverse-countries/

Cardil, A., de-Miguel, S., Silva C.A., Reich, P.B., Calkin, D., Brancalion, P.H.S.,... Liang, J. (2020). Recent Deforestation Drove the Spike in Amazonian Fires. *Environmental Research Letters*, *15*, 121103. doi:10.1088/1748-9326/abcac7.

CAT (2020, December). Paris Agreement Turning Point: Wave of Net Zero Targets Reduces Warming Estimate to 2.1 °C in 2100. All Eyes on 2030 Targets. Berlin: New Climate Institute and Climate Analytics.

Chiaretti, D. (2021, February 5). "Savanização da Amazônia já está ocorrendo", diz Nobre. *Valor Econômico*. Retrieved from https://valor.globo.com/brasil/noticia/2021/02/05/savanizacao-da-amazonia-esta-mais-proxima-diz-nobre.ghtml

De Faria, B.L., Brando, P.M., Macedo, M.N., Panday, P.K., Soares-Filho, B.S., & Coe, M.T. (2017). Current and Future Patterns of Fire-Induced Forest Degradation in Amazonia. *Environmental Research Letters*, *12*(9), 095005. doi:10.1088/1748-9326/aa69ce

Dryzek, J.S., & Pickering, J. (2019). *The Politics of the Anthropocene*. Oxford: Oxford University Press.

Duffy, P.B., Brando, P., Asner, G.P., & Field, C.B. (2015). Projections of Future Meteorological Drought and Wet Periods in the Amazon. *Proceedings of the National Academy of Sciences*, *112*(43), 13172–13177. doi:10.1073/pnas.1421010112

Exbrayat, J.-F., Liu, Y.Y., & Williams, M. (2017). Impact of Deforestation and Climate on the Amazon Basin's Above-Ground Biomass During 1993–2012. *Scientific Reports*, 7(1), 15615. doi:10.1038/s41598-017-15788-6.

Henley, B.J., & King, A.D. (2017). Trajectories Toward the 1.5 °C Paris Target: Modulation by the Interdecadal Pacific Oscillation. *Geophysical Research Letters*, *44*(9), 4256–4262. doi:10.1002/2017GL073480

INPE (2019). Metodologia utilizada nos projetos PRODES e DETER. Retrieved from http://www.obt.inpe.br/OBT/assuntos/programas/amazonia/prodes/pdfs/Metodologia_Prodes_Deter_revisada.pdf/view

IPCC (2018). Summary for Policymakers. In V. Masson-Delmotte, H.-O. Pörtner, J. Skea, P. Zhai, D. Roberts, P. R. Shukla, ... T. Waterfield (Eds.), *Global Warming of 1.5 °C: An IPCC Special Report on the Impacts of Global Warming of 1.5 °C above Pre-Industrial Levels and Related Global Greenhouse Gas Emission Pathways, in the Context of Strengthening the Global Response to the Threat of Climate Change, Sustainable Development, and Efforts to Eradicate Poverty*. Geneva: World Meteorological Organization.

Kormos, C., Mackey, B., Mittermeier, R., & Young, V. (2020, March 20). Primary Forests: A Priotity Nature-Based Solution. *IUCN – Crossroads Blog*. Retrieved from https://www.iucn.org/crossroads-blog/202003/primary-forests-a-priority-nature-based-solution

Lenton, T.M. (2019). Biodiversity and Global Change: From Creator to Victim. In P. Dasgupta, P.H. Raven, & A.L. McIvor (Eds.), *Biological Extinction: New Perspectives* (pp. 34–79). Cambridge: Cambridge University Press.

Lovejoy, T.E., & Nobre, C. (2018). Amazon Tipping Point. *Science Advances*, *4*(2), eaat2340. doi:10.1126/sciadv.aat2340

Malhi, Y., Gardner, T.A., Goldsmith, G.R., Silman, M.R., & Zelazowski, P. (2014). Tropical Forests in the Anthropocene. *Annual Review of Environment and Resources*, *39*, 125–159. doi:10.1146/annurev-environ-030713-155141

Milkoreit, M., Hodbod, J., Baggio, J., Benessaiah, K., Calderón-Contreras, R., Donges, J.F.,… Werners, S.E. (2018). Defining Tipping Points for Social-Ecological Systems Scholarship – An Interdisciplinary Literature Review. *Environmental Research Letters*, *13*, 033005. doi:10.1088/1748-9326/aaaa75.

NOAA (n. d.). Climate at a Glance. Retrieved from https://www.ncdc.noaa.gov/cag/global/time-series/globe/land/ann/12/1880–2020

Nobre, C.A., Sampaio, G., Borma, L.S., Castilla-Rubio, J.C., Silva, J.S., & Cardoso, M. (2016). Land-Use and Climate Change Risks in the Amazon and the Need of a Novel Sustainable Development Paradigm. *Proceedings of the National Academy of Sciences of the United States of America*, *113*(39), 10759–10768. doi:10.1073/pnas.1605516113

Pereira, J.C., & Saramago. A. (2020). Introduction: Embracing Non-Human Nature in World Politics. In J.C. Pereira, & A. Saramago (Eds.), *Non-Human Nature in World Politics: Theory and Practice* (pp. 1–9). Cham: Springer.

Pereira, J.C., & Viola, E. (2018). Catastrophic Climate Change and Forest Tipping Points: Blind Spots in International Politics and Policy. *Global Policy*, *9*(4), 513–524. doi:10.1111/1758-5899.12578

Pereira, J.C., & Viola, E. (2019). Catastrophic Climate Risk and Brazilian Amazonian Politics and Policies: A New Research Agenda. *Global Environmental Politics*, *19*(2), 93–103. doi:10.1162/glep_a_00499

Pereira, J.C., & Viola, E. (2020). Close to a Tipping Point? The Amazon and the Challenge of Sustainable Development Under Growing Climate Pressures. *Journal of Latin American Studies*, *52*(3), 467–494. doi:10.1017/S0022216X20000577.

RAISG (2016). *Amazonía 2016 – áreas protegidas y territorios indígenas*. São Paulo: ISA – Instituto Socioambiental.

RAISG (2020). *Amazonía bajo presión*. São Paulo: ISA – Instituto Socioambiental.

Rammig, A., Jupp, T., Thonicke, K., Tietjen, B., Heinke, J., Ostberg, S.,... Cox, P. (2010). Estimating the Risk of Amazonian Forest Dieback. *New Phytologist*, *187*(3), 694–706. doi:10.1111/j.1469-8137.2010.03318.x

Secretariat of the Convention on Biological Diversity (2020). *Global Biodiversity Outlook 5: Summary for Policymakers*. Montreal: Secretariat of the Convention on Biological Diversity.

Staal, A., Fetzer, I., Wang-Erlandsson, L., Bosmans, J.H.C., Dekker, S.C., van Nes, E.H.,... Tuinenburg, O.A. (2020). Hysteresis of Tropical Forests in the 21st Century. *Nature Communications*, *11*(1), 4978. doi:10.1038/s41467-020-18728-7

Viola, E., & Franchini, M. (2018). *Brazil and Climate Change: Beyond the Amazon*. New York: Routledge.

Warren, R., Price, J., VanDerWal, J., Cornelius, S., & Sohl, H. (2018). The Implications of the United Nations Paris Agreement on Climate Change for Globally Significant Biodiversity Areas. *Climatic Change*, *147*(3), 395–409. doi:10.1007/s10584-018-2158-6

The World Bank (2019, May 22). Why the Amazon's Biodiversity is Critical for the Globe: An Interview with Thomas Lovejoy. Retrieved from https://www.worldbank.org/en/news/feature/2019/05/22/why-the-amazons-biodiversity-is-critical-for-the-globe

WMO (2021, January 15). 2020 Was One of Three Warmest Years on Record. Retrieved from https://public.wmo.int/en/media/press-release/2020-was-one-of-three-warmest-years-record

Zemp, D.C., Schleussner, C.F., Barbosa, H.M.J., & Rammig, A. (2017). Deforestation Effects on Amazon Forest Resilience. *Geophysical Research Letters*, *44*(12), 6182–6190. doi:10.1002/2017GL072955

2 Brazil: The Exterminator of the Future*

The future of the Amazon will mostly depend on Brazil. In addition to the fact that more than 60% of the forest lies within its borders, Brazil is a relevant player in global governance and has significant institutional, scientific and financial capacity compared with its neighboring Amazonian countries. Brazil, the planet's most biodiverse country, once accounted for the highest deforestation rates in the world; in the 1990s and early 2000s, extremely high levels of deforestation – driven by illegal logging, cattle ranching and soybean farming – severely impacted upon the Amazon's flora and fauna. However, in the second half of the 2000s, the country broke from decades of uncontrolled deforestation. Between 2004 and 2012, the Brazilian Amazonian deforestation rate declined by more than 80%. This positive trend has nevertheless been reversed since 2013. During the 2013–2015 and 2016–2019 periods, the average annual deforestation rates in the Amazon have been respectively 25% and 76% higher than the 2012 historical minimum (nearly 4,500 km^2). In 2019, deforestation in the region rose by 30% compared with the previous year, reaching nearly 10,000 km^2 (Appendix A, Figure A.1, PRODES deforestation data).[1]

What political, economic and social factors underlie the trajectory of Brazilian Amazonian deforestation rates of the recent past and present? In the following pages, we provide the answer to this question by analyzing the politics of and policies for the region throughout the period 2005 to 2019. In addition, considering, as already seen, that preventing the Amazon from tipping into a savanna depends not only on good governance on a national level, but also on collective action on a global scale, we examine the Brazilian diplomatic performance at the UNFCCC and CBD.[2] We address part of the first and second

*This chapter draws partly on Pereira and Viola (2019, 2020b) and Pereira and Viola (forthcoming).

administrations of Lula da Silva (2005–2010),[3] Dilma Rousseff's first and second administrations until her impeachment (2011–2016), Michel Temer's administration (2016–2018) and the first year of Jair Bolsonaro's administration (2019).

On the basis of Amazonian deforestation trajectories of the past 14 years, and national political, economic and social dynamics, the discussion is divided into four distinct periods: 2005–2010, 2011–2015, 2016–2018 and 2019. The first period is characterized by the political rise of the country's pro-environmental forces and growing public attention to environmental sustainability issues, both favored by a political and economically stable national context. Between those years, the quality of forest governance in the Amazon improved dramatically. On a foreign level, Brazilian diplomacy played an important role in the design of both the Copenhagen Accord and the Nagoya Protocol. In sharp contrast, the second period is marked by deteriorating economic conditions, increasing political and social instability, decreased political and public attention to issues pertaining to environmental sustainability and the rise of anti-environmentalist forces. Accordingly, between 2011 and 2015, Brazil's foreign policy activism declined. The third period is characterized by a deepening of political instability and the consolidation of the power of the country's anti-environmentalist forces. It culminated in the election of a president who explicitly rejects environmental science and the imperatives of sustainability. Consequently, 2019 witnessed an abrupt shift in Brazilian environmental policy. The resilience of the Amazon is at risk.

2.1 2005–2010: Improvement of Forest Governance

In Brazil, the second half of the 2000s was a period of great stability at all levels, with significant improvements in economic conditions, successful social policies and broad public support for the government. In this favorable context, political and public attention to environmental issues grew in the country, and deforestation control became a political priority for the federal government through the vigorous action of notable Brazilian environmentalists Marina Silva and Carlos Minc, who served as ministers of the environment in Lula da Silva's administrations. Both mobilized a vast coalition for forest protection that resulted in a profound shift in the country's Amazonian politics and policies (Pereira & Viola, 2020b).

During this period, (a) new national parks and conservation units were created by the federal government and the Amazonian state governments (Viola & Franchini, 2018), and a systematic conservation planning[4] approach was adopted to inform biodiversity conservation decisions (Fonseca & Venticinque, 2018); (b) the federal state enhanced its capacity to enforce legislation in the Amazon region through both the coordinated effort of monitoring, inspection and repression by multiple agencies (e.g., the Brazilian Institute of Environment and Renewable Natural Resources – IBAMA, the Chico Mendes Institute for Biodiversity Conservation – ICMBio, both attached to the Ministry of Environment – MMA, and the federal police) and closer collaboration with the various governments of the Amazonian states (Viola & Franchini, 2018); (c) a program (*Programa Bolsa Floresta*) to reward populations making a formal commitment to zero deforestation was established by the state of Amazonas (FAS, n. d.); (d) domestic and international ENGOs and retailers and traders (e.g., McDonald's and Cargill) successfully pressured the Brazilian government and several stakeholders into signing an important meat and soy moratorium opposing the consumption of beef and soybeans produced in deforested areas; and (e) an important fund, the Amazon Fund, was created to raise donations for activities aimed at the prevention, monitoring and combating of deforestation in the Amazon, and the promotion of the conservation and sustainable use of the forest. In 2009 and 2010, Brazil signed finance agreements with Norway and Germany, which were the fund's major donors until its suspension in 2019, as shall be seen later in this chapter. In addition, ENGOs and the scientific community increased their influence on the national media and successfully promoted public awareness of the problem of deforestation in the Amazon (Viola & Franchini, 2018).

Deforestation control in the country resulted in a dramatic reduction in national GHG emissions – more than 40% between 2005 and 2012 (SEEG, 2019) – leading to a shift by the Brazilian diplomacy in the UNFCCC negotiations. Until the second half of the 2000s, fearing any potential foreign interference in the Amazon or international questioning of the economic use of the forest that could undermine the country's sovereignty over the region, Brazil was unwilling to accept emissions targets and refused to acknowledge forests as subjects of international regulation. However, in 2006, during the twelfth Conference of the Paris (COP12) in Nairobi, Brazil proposed an international fund to support deforestation avoidance schemes and, in 2009, during COP15 in Copenhagen, presented one of the most ambitious mitigation targets among the developing countries (Viola & Franchini, 2018).

Since the control of deforestation became an opportunity for accessing international climate financing, the Amazon state governors pressured the Brazilian delegation to the Copenhagen meeting to accept the inclusion of the "Reducing emissions from deforestation and forest degradation and the role of conservation, sustainable management of forests and enhancement of forest carbon stocks in developing countries" (REDD+) international framework into the market mechanisms of the Kyoto Protocol. The Brazilian government and the country's diplomats were also pressured by powerful exporting corporations not only to support REDD+ at the COP meeting, but also to commit to a national voluntary climate change mitigation target, a demand resulting from the passing of the Waxman-Markey bill in the US House of Representatives in 2009, which included tariffs on imports from carbon tax–free countries. In a context in which the Amazon, contrary to what it had always been, was an asset for Brazilian negotiators and deforestation control a source of soft power with low economic and political costs, the influence of the Amazon state governors and powerful corporations, alongside the growing political power of the MMA and the candidacy of former environmental minister Marina Silva to the 2010 presidential elections, helped create the necessary momentum for significant climate policy developments in the country. In addition to supporting the REDD+ mechanism, the government established Brazil's first national plan for climate change in 2008; the following year, it sanctioned a national climate law; in the beginning of 2010, Brazil submitted to the UNFCCC a voluntary emissions reduction commitment of 36%–39% from business as usual by 2020; in the same year, the government regulated five sectoral mitigation plans, which were included in the Copenhagen pledge. Among those plans was the Action Plan for the Prevention and Control of Deforestation in the Legal Amazon (PPCDam) – whose aim was to achieve an 80% reduction in Amazonian deforestation by 2020, a target that accounted for nearly 55% of the Copenhagen voluntary commitment. Accordingly, by the end of this period, the country presented itself as a climate leader and global role model in climate change mitigation governance. Nevertheless, in spite of all positive developments, at the Copenhagen COP Brazil maintained its historical advocacy of the right to development over environmental protection and, in alliance with China, India and South Africa – the country's historical allies in the UNFCCC negotiations – refused a legally binding agreement containing compulsory emissions reduction targets for developing countries (Viola & Franchini, 2018). Behind closed doors, with the US, China, India and South Africa, Brazil helped negotiate the final text of the Copenhagen Accord (Hochstetler, 2012), which laid the foundations for the bottom-up approach that paved the way for the adoption of the Paris Climate Agreement in 2015.

Regarding the protection of biodiversity, this period witnessed a significant slowdown in the loss of forest habitats in the Amazon as a result of the country's policies targeting multiple drivers of deforestation (Secretariat of the Convention on Biological Diversity, 2014). In addition, in line with CBD's provisions, an action plan for the implementation of the national biodiversity policy (established in 2002) was formulated (Ministério do Meio Ambiente, 2006a) and national biodiversity targets for 2010 – including an ambitious goal of increasing the coverage of conservation units to at least 30% of the Amazon – were established in 2006 (Ministério do Meio Ambiente, 2006b); in 2007, the government also launched the national policy for the sustainable development of traditional peoples and communities.[5]

On an international level, Brazil acted cooperatively and played a leadership role in discussions toward an international protocol on access and benefit-sharing (ABS) of genetic resources – the Nagoya Protocol. In 2006, at COP8, together with Malaysia, the country led the developing world and successfully pressured developed countries to agree on a deadline for concluding negotiations (Nijar & Gan, 2012). Aligned with the like-minded megadiverse countries, Brazil advocated a legally binding protocol, with an encompassing scope including derivatives and extracts, which balanced new international access standards (including subjecting access to traditional knowledge associated with genetic resources to prior informed consent and mutually agreed terms) with strong compliance measures (including an obligatory requirement to disclose the origin of genetic resources/traditional knowledge in patent applications and checkpoints in user countries) underpinned by sanctions; in addition, provisions of international agreements such as the Agreement on Trade-Related Aspects of Intellectual Property Rights (TRIPS) of the World Trade Organization should, in those countries' view, be harmonized with the CBD's protocol on ABS (Cabrera Medaglia, 2015; Nijar & Gan, 2012; Wallbott, Wolff, & Pożarowska, 2014).

Brazil was also an active player during discussions regarding the creation of the Intergovernmental Science-Policy Platform on Biodiversity and Ecosystem Services (IPBES) – an independent intergovernmental body which assesses the state of the planet's biodiversity and ecosystem services. It emphasized, nevertheless, that the panel's aim should not be to prescribe policies, as this would affect the country's sovereignty over its natural resources (Lovejoy & Inoue, 2012).

At the decisive COP10 in Nagoya, Brazil often acted as a mediator between different points of view and interests, and facilitated agreements. In contrast to the position it had displayed during negotiations for the 2010 biodiversity target in 2002 – when the country's negotiators advocated that each party should establish its own priorities regarding conservation, eventually weakening the program agreed at COP6, which lacked concrete, quantifiable goals (Couto, 2012) – in 2010 Brazil, although initially reluctant to the establishment of any clear deadline, ended up working with Switzerland and the African Group to propose the formulation that by 2020, the rate of loss of all natural habitats, including forests, would be at least halved (Orsini & Diallo, 2015). However, to pressure the developed world, Brazilian negotiators, aligned with the G77 and China, made the adoption of CBD's new strategic plan for the 2011–2020 period dependent on the conclusion of the protocol on ABS (Mittermeier et al., 2010). As noted by Wallbott et al. (2014, p. 48),

> Northern countries could hardly reject the suggestion if they wanted to avoid blame for not halting the continuing loss of biodiversity (…). Furthermore, another failure of UN negotiations after the 2009 climate summit in Copenhagen would have put the system of multilateral environmental governance as a whole in crisis.

Nevertheless, in spite of its alliance with the developing world, particularly with the like-minded megadiverse countries, Brazil worked closely with the European Union at the meeting. The Nagoya Protocol is, to a very significant extent, the product of negotiation between those two parties. The protocol's final draft was concluded behind closed doors by Brazil, the European Union, Japan, Norway and the African Group – a procedure that was highly criticized for its lack of transparency (IISD, 2010) and which angered some of Brazil's traditional allies within the CBD (Nijar, 2011). COP10 marked the beginning of a shift by Brazil from the positions it held in common with the like-minded megadiverse countries (Nijar, 2011); in fact, the legal text approved in Nagoya reflected the European Union's objectives to a high degree (Groen, 2019). To understand such a shift, and some of the subsequent developments that will be addressed in the next subsection, we need to look at the Brazilian domestic context.

The ABS issue was a relevant one for several ministries in Brazil – environment, science and technology, industry and agriculture in particular. These had different and, in some cases, opposing views and interests regarding a potential international protocol on the matter. The MMA supported an international protocol for obvious reasons related to the protection of biodiversity and traditional knowledge. On the contrary, the Ministry of Science and Technology, fearing that an international treaty could hinder access to genetic resources for research purposes, had a more defensive view of the potential protocol. The ministry also viewed Brazil's alliance with the like-minded megadiverse countries as a weakness that could compromise the country's biotechnological development. The Ministry of Industry feared potential impediments to research for commercial uses and further obligations regarding the sharing of benefits. Finally, the Minister of Agriculture also had, for reasons that will be detailed in the next subsection, reservations regarding an international protocol on the issue (Moreira, 2016).

The position of the MMA prevailed. During this period, the ministry had strengthened its influence over the core of the federal government. Moreover, and perhaps most importantly, among all ministries, the MMA had the greatest knowledge of the complex issue of biodiversity and the CBD's negotiations. At the time, the Secretariat of Biodiversity and Forests was headed by Bráulio Dias, an experienced biologist who had participated in all CBD's COPs as a member of the Brazilian delegation, and who would become CBD's Executive Secretary in 2012. Dias had a key influence over the process. It should be noted, nevertheless, that at COP10 the Minister of Environment was Izabella Teixeira, a qualified technocrat with no history of environmental militancy, who had taken office only a few months earlier. Unlike her predecessors, especially Marina Silva, who usually displayed a somewhat offensive stance during discussions on ABS, Teixeira was more pragmatic. In addition, the Brazilian delegation to Nagoya had been instructed by the country's Civil House to conclude the agreement, and efforts were thus directed accordingly (Moreira, 2016).

Until the Nagoya COP, Brazil had acted like a Southern, developing country, focusing on preventing national biodiversity and traditional knowledge from being acquired illegally. However, the provider–user distinction was being blurred, and attempts were being made in the country to reconcile both roles, as Brazil was no longer only a supplier, but also, and increasingly, a user of genetic resources. Brazilian stakeholders were emphasizing the need for a stronger national biotechnology sector, seeking to make the most of the country's genetic resources, rather than thinking exclusively in terms of biopiracy (Filoche, 2013) – in 2007, the Brazilian government had even launched a national policy for biotechnology development.[6] In reality, the country's interests were no longer fundamentally different from those that guided developed countries; Brazil would not benefit from radically changing the rules of global governance in that domain (Muzaka & Serrano, 2019).

2.2 2011–2015: Stagnation and Regression

During this period, Brazil lived through turbulent times that negatively impacted forest governance and the country's international performance.

Contrary to what had happened in the previous years, environmental protection issues played a peripheral role in the political agenda of Lula da Silva's successor, Dilma Rousseff, whose administration prioritized short-term economic growth considerations. The MMA, which continued to be headed by Izabella Teixeira, resigned itself to the vision and orientation followed by the new administration. In 2011, the ministry supported a reform to the Brazilian Forest Code that largely reduced environmental protections (Novaes & de França Souza, 2013) and marked the beginning of the rise of the country's anti-environmentalist forces, leading the environmental movement to break the alliance it had maintained with the Workers' Party (PT) since 2003. The reform was promoted by the Agribusiness Parliamentary Front (FPA), known as the ruralist caucus, which accounted for 28% of the representatives in the 2011–2014 legislature, 7% more than in the previous one. During the process, it became evident that the agribusiness coalition was able to exert significant power over the executive and legislative branches of government. Modern and competitive, with remarkable organizational robustness and accounting for nearly half of Brazil's exports, thus playing a major role in the country's trade balance, the agribusiness sector would further increase its political power over the following years. Their support would become pivotal in passing legislation – a fact that, as shall be seen, allowed the FPA to lead an offensive policy against the country's environmental legislation (Pereira & Viola, 2019; Pereira & Viola, 2020b).

It should also be noted that, during this period, as a result of both the failure of the Waxman-Markey bill to pass in the American Senate and the defeat of the Democratic party in the 2010 mid-term elections – which dramatically reduced any possibility of passing climate legislation in the United States – the Brazilian business sector, which had been an active player in favor of environmental and climate action in the previous years, ceased to be a driving force in environmental policy (Viola & Franchini, 2018).

There were, however, some positive developments in the early 2010s. In 2011, the Brazilian government established a program (*Bolsa Verde)* aimed at reducing rural poverty by benefiting those who lived within conservation areas in conditions of extreme poverty and who were committed to nature conservation.[7] In 2012, the government created a national policy for the territorial and environmental management of indigenous lands.[8] In 2013, building on the CBD's Aichi Targets Brazil released national biodiversity targets for 2020. These included halving the rate of loss of native habitats, significantly reducing the risk of extinction of threatened species, enhancing the resilience of ecosystems and increasing the contribution of biodiversity to carbon stocks (Ministério do Meio Ambiente, 2013). Nevertheless, the actual impact of any laws, resolutions and programs on biodiversity and deforestation "depends as much on the level of enforcement and effectiveness, as it does on the ambitions" (UNEP-WCMC, 2014, p. 2), and the quality of environmental governance and attention to environmental issues in the country would decrease dramatically over the following years.

By the middle of this period, Brazil had begun to struggle with rising inflation, falling commodity prices, deteriorating public services, a huge corruption scandal (known as Operation Car Wash) involving politicians, the country's state-run oil firm Petrobras and several national infrastructure companies such as Odebrecht, as well as massive street demonstrations that culminated in a severe economic recession and a cycle of political instability that would later lead to the impeachment of the president. Economic and political turmoil reinforced the strength of the FPA and the Rousseff administration's neglect of environmental protection issues. Political and public attention to the environment in the country declined dramatically – in contrast to what had happened in the campaign discussions for the 2010 presidential elections, in which the environmental and climate agendas were brought to the spotlight, those of 2014 were dominated by issues such as unemployment, economic growth, public spending, inflation control and corruption; in spite of the fact that Marina Silva ran once more for the country's presidency, the environment was largely overlooked (Pereira, 2019; Pereira & Viola, 2019; Pereira & Viola, 2020b).

The deteriorating state of Brazil's economic, social and political situation almost cost President Rousseff her re-election. In a climate of instability, for her second mandate, the president chose Kátia Abreu, who at the time was the president of the Brazilian Confederation of Agriculture and Livestock (CNA) – the largest representative of the country's rural producers before the Brazilian Congress – and the leading figure of the FPA, to head the Ministry of Agriculture. Abreu was known as "Miss Deforestation" due to her anti-environmentalist positions (Kröger, 2017).

Within this context, it should come as no surprise that during Rousseff's first term, the creation of new conservation units was interrupted and, in some cases, the size of existing protected areas was reduced (Alencastro, 2014; Imazon, 2012); by the end of her second mandate, the president had created only 15 federal conservation units comprising approximately 37,000 km^2 – seven of those were created in the lead-up to the second round of the 2014 presidential elections, a move that some attributed to an attempt to attract the vote of Marina Silva's electorate (in the first round, the former environmental minister accounted for 24% of the popular vote) (ISA, 2016). This period thus saw the triumph of a development paradigm according to which forest protected areas are obstacles to national progress. In comparison with previous governments, the contribution of the Dilma administration to the effort of protecting ecologically relevant areas in the country was quite insignificant – putting numbers into perspective, during President Lula da Silva and President Fernando Henrique Cardoso's mandates, 77 and 81 conservations units, comprising nearly 267,000 km^2 and 208,000 km^2 respectively, had been created (ISA, 2016).

Moreover, effectiveness in the management of existing protected areas, including within the Amazon, was far from ideal (Prates & Irving, 2015) – a problem that would persist over the years, aggravated by the weakening of the institutional capacity of IBAMA and ICMBio as a result of the worsening of Brazil's fiscal situation, which led, and would continue to lead, to significant reductions in financial support for those agencies (Pereira & Viola, 2020b).

Finally, the weakened institutional capacity of the MMA, alongside the fact that the new Forest Code included an amnesty for past environmental crimes, thus contributing to a generalized sense of impunity, encouraged organized criminal groups and vulnerable rural communities hit by fast-growing unemployment to deforest illegally (Pereira & Viola, 2020b).

In sum, although Brazil has capable institutions and proper environmental laws, it does not execute them adequately or create the conditions or instruments to make their enforcement possible.

Within the UNFCCC, Brazil regressed to its traditional low level of commitment to climate change mitigation, based on a radical interpretation of the principle of "common but differentiated responsibilities and respective capabilities" (CBDRRC) that traces the historical responsibilities for climate change back to the Industrial

Revolution. The control of deforestation was used by Brazilian negotiators to argue that Brazil had already gone beyond its mitigation obligations (Federative Republic of Brazil, 2015). Given the declining political influence of the MMA, the more nationalistic standing of President Rousseff's foreign policy and the low public and political attention paid to environmental issues in the country, the conservative vision of the Ministry of Foreign Affairs, to whom the international environmental agenda constitutes a threat to economic growth and national sovereignty over territory and natural resources, prevailed (Viola & Franchini, 2018).

On the road to Paris, Brazil, pressured by the developed world and seeking to exempt major developing countries from assuming binding emissions-reduction commitments while portraying itself as an active player, forwarded an idea that would influence subsequent negotiations – the "concentric circles" proposal, which is simply a corollary to the CBDRRC principle (Albuquerque, 2019). The "concentric differentiation" approach proposed by the country divided parties into three circles: the center with developed and major emerging economies, which should commit to economy-wide absolute GHG emissions reduction targets; the intermediate ring containing developing countries, which should adopt relative economy-wide targets; and the outside ring with the poorest countries, which would not have any obligations. All countries would be expected to move toward the inner circle over time and in accordance with their CBDRRC. The proposal, which reaffirmed Brazil's foreign policy tradition of dividing the globe into rich and poor countries, was, nevertheless, celebrated as a creative and progressive one by the country's diplomacy (Viola & Franchini, 2018).

In 2015, under the Paris Climate Agreement, Brazil pledged to reduce GHG emissions by 43% below 2005 levels in 2030, not contingent upon international support; no peak emissions year was defined. CAT (2017) rates Brazil's nationally determined contribution (NDC) as "insufficient" – the country's level of climate commitment is inconsistent with limiting global warming to below 2 °C, "unless other countries make much deeper reductions and comparably greater effort"; although "Brazil is one of the few developing countries that has put forward absolute emission reduction targets in their NDC (…) the CAT finds those target levels to be at the least ambitious end of a fair contribution to global mitigation" (CAT, 2017). In fact,

> while nominal reduction targets appear to be challenging and ambitious at first glance, after taking into account that the base year for the NDC targets (2005) was a year of particularly high

> emissions, the real target represents very little effort beyond current ambition levels. (…) the NDC effectively translates to a decrease of only 7% in emissions incl. LULUCF [Land Use, Land-Use Change and Forestry] below 2012 levels by 2030 (CAT, 2019b).

According to CAT (2019b), if all counties followed the Brazilian level of climate change mitigation ambition, the planet would warm by 2–3 °C by the end of the century. Others consider Brazil's NDC even less ambitious. For example, Du Pont and Meinshausen (2018, from the Paris Equity Check) suggest that if taken as a benchmark by other countries, the Brazilian pledge would lead to warming of 3.7 °C by 2100. As shall be seen below, by the end of 2019, Brazil was not even on track to meet its climate change mitigation pledges.

Regarding the CBD, President Rousseff sent the Nagoya Protocol to the Congress for ratification in 2012. However, as the protocol referred to the states' legal framework for several specifications, Brazil would need adequate legislation on ABS (Medeiros & Albuquerque, 2015). Motivated precisely by discussions for the Nagoya Protocol, several actors were already calling for a reform of the country's regime on the matter since the previous year. In force since 2001, the regime,[9] which was designed without the participation of the Brazilian civil society, was highly restrictive and consequently mostly impracticable for researchers – thus impeding critical research for the protection of biodiversity – and companies, both national and foreign (Da Silva Bolzani, 2017; Filoche, 2013). It was also failing to protect biodiversity and traditional knowledge from biopiracy, and to ensure the sharing of benefits with indigenous communities (Gonçalves de Andrade, 2017). In addition, the FPA in particular made its support of the ratification of the Nagoya Protocol partly dependent on the implementation of new ABS legislation that ensured legal certainty and safeguarded the sector's interests. Moreover, since 2010, IBAMA had been applying heavy fines to companies, research centers on biotechnology and universities for using genetic resources and associated traditional knowledge without permission. Those fines represented real monetary losses for large, important national companies, especially in the pharmaceutical and cosmetics sectors, and mobilized their representatives to engage in the debate on the country's regime on ABS (Moreira, 2016).

The new Brazilian ABS law[10] was signed in 2015 following a complex negotiating process among different ministries and diverse sectors of civil society (academia, business sector and holders of associated traditional knowledge) with divergent visions and interests. The law was highly criticized by organizations representing small farmers, indigenous peoples and traditional communities, which accused the federal government and the country's agribusiness and industrial sectors of excluding their representatives from its drafting process. In fact, the new law does not protect or ensure the fair and equitable sharing of benefits with those actors, instead favoring the pharmaceutical, cosmetics and agribusiness sectors. It contains several possibilities for users to be exempted from obtaining prior informed consent from and sharing benefits with indigenous peoples and traditional communities (e.g., in the case of accessing genetic resources for agriculture and food purposes); the new law also amnesties companies that were fined before its coming into force. In spite of the wording of the Nagoya Protocol, which explicitly implies greater protection for those actors, their legal position was significantly weakened in Brazil (Gonçalves de Andrade, 2017; ISA, 2015a; 2015b; see also Moreira, Porro, & Silva, 2017). The debilitated MMA gave in to the ministries and sectors which focused exclusively on the economic value of genetic resources and the associated traditional knowledge, to the detriment of the rights of indigenous and traditional communities.

Criticisms of the law also came from the country's academia, which were disappointed by the fact that it did not recognize non-commercial research as a special area that should be facilitated, thus inhibiting basic and applied research as well as international cooperation by including "technically impracticable" requirements – involving, for example, high operational costs and legal barriers affecting access to Brazilian biodiversity by foreign researchers (Alves et al., 2018).

The process of designing the new Brazilian ABS law exposed once more the country's inability to deal effectively with strategic issues of national interest demanding a long-term vision and involving powerful economic sectors.

The FPA was a vocal actor in national discussions on the new ABS law and the ratification by the country of the Nagoya Protocol. In the FPA's view, the Brazilian ratification of the protocol could hinder the country's access to foreign genetic resources for research and breeding purposes, and transform the country into one of the world's largest royalty payers, as several genetic resources exploited by the country's agribusiness sector, such as those of soya and corn, are exogenous to Brazil. For the ruralists, the Nagoya Protocol should not include agricultural biodiversity; Brazil's priority should be to work toward broadening the scope of the International Treaty on Plant Genetic Resources for Food and Agriculture (TPGRFA) of the United

Nations Food and Agriculture Organization (adopted in 2001) – a treaty declaring that 64 of the planet's most important crops will comprise a pool of genetic resources that are accessible to everyone – so that it included all plant species used in Brazilian agriculture (Escobar, 2014; Globo Rural, 2013; ISA 2014). This position was shared by Kátia Abreu (Magalhães, 2013), who, as previously stated, became Minister of Agriculture in 2015.

The FPA's arguments, which have been maintained until the present day, ignore the facts that the protocol has no retroactive effects, which means, for example, that it does not affect the species varieties already used in Brazilian agriculture. In addition, the genetic improvement of soy – a major concern for the FPA – conducted in Brazil is mainly based on genetic resources that are already available in *ex situ* collections maintained by Brazilian institutions. Moreover, conditions for accessing genetic resources are determined by national legislations, not by the Nagoya Protocol. Even if Brazil does not ratify the protocol, it will have to comply with the ABS legislation of the origin countries (ISA, 2014).

Due to the FPA's resistance, Brazil – a key player in the Nagoya negotiations – participated as a mere observer at the first meeting of the parties to the protocol, held during CBD's COP12 in 2014, thus weakening its position within the convention, especially with regard to negotiations for the operationalization of the protocol. Being an observer, the Brazilian diplomacy would find it harder to influence the process to the country's benefit. In addition, by not ratifying the protocol, Brazil, which is potentially the most affected country in the world by the uncontrolled withdrawal of natural resources from its territory, loses a critical opportunity to protect its biodiversity and associated traditional knowledge from biopiracy, and to combat rural poverty, which is, as we have seen, one of the drivers of forest degradation and biodiversity loss in the country.

2.3 2016–2018: Aggravation of the Predatory Trend

This period was first marked by the impeachment of President Dilma Rousseff, who lost control of her coalition in early 2016 amid a severe economic recession, mounting evidence of deep corruption within the PT and charges of fiddling government accounts. When discussions on the impeachment began, the FPA soon saw in the process an opportunity to further strengthen its political power and advance its anti-environmentalist agenda. In March 2016, the agribusiness coalition – accounting for 39% of the deputies in the national Congress at the time

– declared its full support for Rousseff's impeachment; in April, its leaders discussed the group's main concerns and positions with Brazilian Vice-President Michel Temer. In August, Rousseff was impeached – unsurprisingly, half of the votes against the President came from the agribusiness caucus – and Temer became Brazil's new President (Pereira & Viola, 2019; Pereira & Viola, 2020b).

During his tenure, Temer faced very low rates of presidential approval (<10%), corruption accusations and two impeachment votes. In return for the support of the FPA to pass his legislative agenda and avoid trial and impeachment, the president met some of the ruralists' demands and took major measures against the Amazon, signing a series of provisional acts, decrees and laws that reduced the size of protected areas and allowed land grabbers to legalize their holdings in the region, suspended the ratification of indigenous lands and forgave farmers and ranchers billions of dollars in environmental fines and debts (Pereira & Viola, 2019; Pereira & Viola, 2020b). In addition, in the months preceding his impeachment votes, Temer released vast sums in pork-barrel allocations (*emendas*) and other expensive concessions to ruralist representatives (Fearnside, 2018).

The situation was further aggravated by the Temer administration's policy to drastically reduce the country's fiscal deficit, with severe cuts in environmental and science spending undermining the MMA's institutional capacity and research on biodiversity which, as already mentioned, is a critical basis of public policies for sustainable development and nature protection (Magnusson et al., 2018; Pereira & Viola, 2019; Pereira & Viola, 2020b). In addition, the *Bolsa Verde* program – which was having a non-negligible positive impact on Amazonian deforestation – was suspended (Scarano & Silva, 2018). Although the new Minister of Environment, José Sarney Filho, was a figure close to the environmental movement, the MMA found itself unable to avert further deterioration of forest protection policies.

By the end of this period, the country's poor environmental performance was evident. For example, only 15.5% of the country's flora – 5,646 species – had been assessed for its conservation status, and approximately half of those species were classified as threatened (Martins, Martinelli, & Loyola, 2018); 85 of those species were located in the Amazon (WWF-Brasil, 2019). Brazil was therefore far from achieving target 2 of the Global Strategy for Plant Conservation (GSPC) of the CBD, which states that by 2020, all parties had to have undertaken risk assessments of their entire known plant species to guide conservation action. Regarding targets 6 and 11, according to which by 2020, at least 75% of production lands would have been being managed sustainably, consistent with the conservation of plant biodiversity, and no species of wild flora would have been being endangered by international trade, results were also unpromising. The area of native ecosystems converted to cropland and pastureland remained high, particularly in the Amazon and the Cerrado, as did the number of species threatened by illegal timber exploitation, especially in the Amazon and the Atlantic Forest (Scarano & Silva, 2018). Progress on target 10, which regards the implementation of effective management plans to prevent new invasions by alien species – major threats to native plants – and the management of important areas for plant diversity that are invaded – was also slow. Very few management plans had been implemented; the national strategy on invasive alien species had not yet been implemented (Dechoum, Sampaio, Ziller, & Zenni, 2018). In addition, despite the fact that 18% of Brazil was a terrestrially protected area,[11] it would be difficult to assert that the country had met Aichi target 11 – which states that by 2020 at least 17% of terrestrial areas are conserved through effectively and equitably managed, ecologically representative and well-connected systems of protected areas – due to the lack of protected areas outside the Amazon and to poor management of existing conservation units, including within the Amazon (Pacheco, Neves, & Fernandes, 2018). It should also be noted that the Amazon was the Brazilian biome with the largest number of events of downsizing, downgrading and degazettement of protected areas (Vieira, Pressey, & Loyola, 2019).

A positive development during this period, and perhaps the only one, was the launching of the Brazilian Platform on Biodiversity and Ecosystem Services (BPBES) in 2017 with "the mission of producing regular assessments of the best available knowledge, by science and other knowledge systems, on biodiversity and ecosystem services issues (…) and to be used to improve the interface between science and policy" (Padgurschi & Joly, 2017, p. 1).

Public neglect of environmental issues persisted, with most Brazilians showing deep concern for the country's unemployment numbers, systemic corruption, poor quality of health care services and rising levels of crime and violence. During campaign discussions for the 2018 general elections, the environmental agenda was once more completely overlooked. In fact, the presidential election was won by Jair Bolsonaro, whose campaign had been characterized by a strong anti-environmentalist discourse (Pereira & Viola, 2019, 2020b).

Within the climate regime, Brazil blocked negotiations on the carbon market mechanisms of the Paris Climate Agreement at COP24 in December 2018. When discussing rules to prevent parties from double counting emissions reductions – that is, to ensure that a country paying another to lower emissions does not count those emissions reductions as part of its mitigation progress – the Brazilian delegation relentlessly insisted that developing countries should be exempted from those rules, a position that earned Brazil harsh criticism (Pereira & Viola, forthcoming).

In the CBD, the country's official position was highly influenced by the Ministries of Agriculture and Industry (Bittencourt, 2018; Dallagnol & Rodrigues, 2016). At COP13, the Brazilian delegation argued that studies on the impacts of living modified organisms on pollinators are inconclusive and suggested deleting reference to insecticides and fungicides on risk assessment procedures for pesticides (IISD, 2016). At COP14, Brazil voted against the monitoring and control of the use of synthetic biology and opposed a moratorium to temporarily ban the release of organisms carrying gene drives (Bensusan, 2018; Bittencourt, 2018).

2.4 2019: Disruption of Environmental Policy

Following years of deterioration of environmental governance in Brazil, the year 2019 started a new phase characterized by climate change skepticism and a clear repudiation of environmental sustainability imperatives by the new government. From the beginning, President Bolsonaro and his government made no secret of their anti-environmentalist agenda. The first year of the new administration was marked by a blatant anti-science discourse that denied the anthropogenic character of climate change and by successive attempts at disrupting Brazil's environmental policy – a stance that contradicted three decades of governmental efforts by different governments to disseminate an image of Brazil as an environmentally-concerned and active country in global environmental fora. Over the past three decades, even when pursuing an insufficient or negligent environmental policy, all Brazilian administrations have never denied the importance of climate change and the protection of biodiversity, and worked toward building an idea of Brazil as an environmental leader. The Bolsonaro administration has explicitly aligned with the also openly anti-environmentalist US Donald Trump administration. Bolsonaro's decision in November 2018 – even before taking office – to withdraw from hosting the UNFCCC's COP25 was the first glimpse of the abrupt shift that the new administration would bring to the country's environmental policy (Pereira & Viola, forthcoming). To understand such a shift, it is important to look first at the drivers of the election of Bolsonaro and at the composition of his government.

What factors explain the election of Bolsonaro to the Brazilian presidency? First, the effects of the economic, political and moral crisis that had been hitting the country since 2013 undermining the credibility of Brazilian political parties. In 2018, Brazilians were urging for political renewal, and many had developed a strong "anti-PT" sentiment due to both the perception that the party led the country to economic decline and the evidence of its involvement in systemic corruption schemes. During his campaign, Bolsonaro took advantage of widespread popular discontent by creating a narrative against the Brazilian political system, thus speaking directly to most voters' main concerns. This narrative was critical in convincing the electorate that Bolsonaro was, among all presidential candidates, the most well-positioned to triumph over Fernando Haddad, PT's candidate. Second, the support of Brazil's business and financial sectors. Early in his campaign, Bolsonaro stated his intention to nominate market-friendly economist Paulo Guedes as Minister of Economy if he was elected, thus becoming the preferred candidate of the country's entrepreneurs and financiers, which feared another left government. Third, Bolsonaro's socially conservative ideas – being for the promotion of family values and against abortion, gay marriage and drug legalization – which earned him the votes of the country's evangelicals, who account for nearly 30% of the Brazilian electorate (Pereira & Viola, forthcoming).

The strategic core of the new government in 2019 was formed by the following ministries. First, a group of market-friendly ministries – economy, agriculture, infrastructure and mines and energy, led respectively by Paulo Guedes, former leader of the FPA Tereza Cristina, engineer Tarcísio Freitas and Admiral Bento Albuquerque. Their strategic vision for the country's development in multiple areas – for example, supporting mining on and agribusiness leasing of indigenous lands, or the construction of large infrastructure in the Amazon region – is at odds with the imperatives of environmental protection. Second, the Ministry of Justice, headed by Sérgio Moro, the federal judge who led Operation Car Wash, for whom the fight against corruption, organized crime and violence should be the ministry's priority for the 2019–2022 legislature. Third, a group of nationalist-nativist ministries – environment, education and foreign affairs. The Minister of Environment, Ricardo Salles, is a lawyer who worked for the Brazilian Rural Society (SRB, an association of rural producers), and is thus close to the ruralists. In December 2018, he was convicted for environmental fraud – during his tenure as Secretary of the Environment of the state of São Paulo, Salles altered an environmental plan in favor of businesses. The

Minister of Education, Abraham Weintraub, and the Minister of Foreign Affairs, Ernesto Araújo, are followers of Olavo de Carvalho, a Brazilian self-proclaimed philosopher who developed "olavism" (*olavismo*), which is a set of ideas, mostly conspiracy theories (e.g., smoking is not harmful for one's health, the climate change agenda is an invention by the Chinese government to weaken the economies of the West), compiled by Olavo himself. Olavists vilify the political left and frequently distort historical and scientific facts. For example, the Minister of Foreign Affairs often publicly denies the anthropogenic character of climate change. It should be noted that Bolsonaro and his three sons – Senator Flávio, Federal Deputy Eduardo and Municipal Representative Carlos – are also olavists. Fourth, and lastly, a group of military-led ministries – science and technology, defense and the presidency's institutional security office – who support a deepening of the state's control over the Amazon through, for example, the construction of new infrastructure in the region (Pereira & Viola, forthcoming). Since the 1970s, when environmental issues became an international political matter, Brazil, particularly the military sector, has feared the possibility of indigenous lands becoming independent or supranational territories administered by the United Nations (Tigre, 2017).

During the first year of the Bolsonaro administration's mandate, several anti-environmentalist measures and actions were taken. For example, the MMA's Secretariat of Climate Change and Forestry was made extinct, and the budget of the ministry suffered severe cuts. Moreover, acting against what Bolsonaro and Minister Salles call "the environmental fines industry", the new government neutralized the role of IBAMA – harshly criticized by the new administration – in imposing fines by creating a regulatory body with the power to revise or quash them (Pereira & Viola, forthcoming). From January to November 2019, the number of environmental fines declined by 25% compared with the same period of the previous year; the total number of environmental fines imposed in 2019 was the lowest in 15 years (Oliveira, 2019). In addition, following public release of deforestation data for the April–June period, the Bolsonaro administration not only exonerated the Director of the INPE – the federal institute that monitors and tracks deforestation in the Amazon – but also opened a public call aimed at selecting a private company to replace the institute in the monitoring of the forest. Another disastrous initiative was the attempt by Salles to alter the destination of the Amazon Fund to indemnify land expropriation. As a result, Norway and Germany suspended their donations to the fund. Other anti-environmentalist actions were the restructuring of the national environmental council

(CONAMA) with a dramatic reduction in the number of representatives of civil society; the approval and relaxation of control of the use of several highly hazardous pesticides, many of them banned in the European Union (Pereira & Viola, forthcoming); and the signing of a presidential provisional measure[12] significantly easing regulations for registering land claims in the Amazon, which could boost land grabs and rural conflict and violence (Ministério Público Federal, 2020). Finally, Bolsonaro also declared its support for the Brazilian Association of Soya Producers' demand for the extinction of the soy moratorium in the Amazon (Caetano, Mendes, & Ramos, 2019) to cover the Chinese demand for soy beans in the context of the US–China trade war (Fuchs et al., 2019).

The Bolsonaro administration's explicit anti-enviromentalist agenda, alongside its aggressive and inflammatory discourse against indigenous communities and their lands – which represent 23% of the Amazon and are important barriers to Amazonian deforestation and biodiversity loss (Lima et al., 2020; Nepstad et al., 2006) – encouraged illicit forest conversion activities and violence against indigenous peoples by land grabbers, petty miners and criminally organized loggers (Pereira & Viola, forthcoming). Governmental rhetoric and action contributed to a generalized climate of impunity in the country and further fed the feeling of empowerment of those who engage in illegal deforestation. The effects of Bolsonaro's attack on the environment profoundly contradicted his campaign promise to fight crime in Brazil (Pereira & Viola, forthcoming).

The consequences for the Amazon were severe. In August and September 2019, large areas of the forest burned in devastating fires. According to the analysis conducted by MAAP (2019), many of the fires followed 2019 deforestation events. The Bolsonaro administration's response to the Amazon crisis led to a global wave of indignation and largely damaged the international image of Brazil. Instead of admitting the gravity of the situation, the Brazilian government downplayed the problem, accused ENGOs of starting the fires, repeatedly denied the global ecological significance of the forest and questioned and refused foreign aid, obsessively reaffirming the country's sovereign right to govern the Amazon and offending the European states that offered to help. For the Brazilian government, global concern over the fires in the region was unreasonable, and the political condemnation of Brazil by the international community was unfair and part of an ideological campaign against the country. This episode further reinforced the Brazilian elites' long-standing territorial anxieties over the integrity of the Amazon (Pereira & Viola, forthcoming).

For President Bolsonaro and Minister Salles, the national policy for the Amazon, the poorest region in Brazil, should prioritize socioeconomic development, and environmental protection restrictions are an obstacle to that development. Accordingly, the government soon revealed its intention to build new infrastructure in the region (e.g., dams, highways and bridges), allow mining on indigenous lands and legalize agribusiness leasing of these territories (Pereira & Viola, forthcoming), an idea that is supported by some communities (Lima et al., 2020). In addition to increase Brazil's revenues, such activities would also contribute to strengthening governmental control over the Amazon. Headed by Salles, the MMA's priority is the country's urban environment, not the country's forests; the minister often stated that environmental problems in Brazil's urban centers (e.g., lack of treated water and excessive waste) were highly overlooked in environmental policy by the past administrations and that the country's major environmental problem is urban, not rural. The minister's policy is supported by the ruralists, who in 2019 accounted for over 50% of the Brazilian Congress. Although a part of the agribusiness coalition feared international boycotts of the country's agricultural products as a consequence of the government's disastrous environmental policy, the more reformist branches of the sector remained very limited (Pereira & Viola, forthcoming).

Public neglect of environmental issues persisted in 2019, and not even the Amazon fires mobilized the necessary support for the protection of the forest. There was no visible improvement in economic conditions and unemployment levels, and corruption investigations were blocked by forces of both the political right and left. Since there are 28 parties represented in the Brazilian Congress and Bolsonaro did not form a coalition, his ministers are limited in their capacity to address some of the country's most pressing problems (Pereira & Viola, forthcoming). In December 2019, Bolsonaro's disapproval rate was at 38% (Barbiéri, 2019), making him the least popular elected president in a first mandate in the history of Brazil.

Unsurprisingly, in 2019, the Amazonian deforestation rate accelerated significantly (Figure A.1) as well as the number of invasions of indigenous territories (Faleiros & Nascimento, 2019). According to CAT (2019a), by the end of 2019, Brazil was on track to miss its 2020 deforestation targets by a large margin, and its economy-wide NDC targets.

Salles's disastrous performance as Minister of Environment earned him the "Exterminator of the Future" award – "a doll wearing a suit and tie and holding a sign over the stump of a felled tree", which was handed to him by a young protester during a public hearing of the Committee on the Environment and Sustainable Development in

Congress in October 2019 (Folha de S. Paulo, 2019). The expression "Exterminator of the Future" had first been used by Former Minister Marina Silva earlier in that year to denounce the anti-environmentalist agenda of the Bolsonaro administration (Gortázar, 2019).

For the UNFCCC COP25, Brazil sent a shrunken delegation led by Salles. For the first time since 1992, accreditation to take part in the official delegation was denied to civil society; in addition, breaking with a 12-year-old tradition, the Bolsonaro administration opted not to have a government booth at the conference. During the meeting, Salles focused mostly on economic development and agribusiness opportunities in the Amazon as well as on the country's (past) good results in forest preservation and share of renewable energy sources in the energy matrix, thus ignoring the recent increase in deforestation and escalating violence against Brazilian traditional populations (Nobrega, 2019; The Brazilian Report, 2019). Aligning with its historical allies, China and India, Brazil opposed placing any obligation on parties to submit more ambitious pledges and refused to engage in any discussion on enhancing their current targets before 2020 without the delivery of finance and support promised by the developed world. Regarding the carbon market mechanisms of the Paris Climate Agreement, Brazil, together with countries such as Australia and India, argued that parties should be able to use old, unspent carbon credits of the Clean Development Mechanism in the new system, which could undermine the system's entire functioning; supported by Australia, the country also continued to oppose rules for preventing double counting of emissions reductions (Newell & Taylor, 2020). Moreover, Brazil was among the countries that opposed the inclusion of the expression "climate emergency" in COP25's final declaration (The Brazilian Report, 2019).

Expectations for the Brazilian participation at CBD's COP15, which was scheduled to be held in 2020 in China but has been postponed due to the COVID-19 pandemic, are very low.

How far could the Bolsonaro administration go in eroding Brazil's environmental policy? This remains an open question. However, it seems certain that the difficult economic situation brought about by the pandemic will bring additional challenges to forest conservation.

Notes

1 It should be noted, however, that, according to a study by Kalamandeen et al. (2018), small-scale forest loss in the Brazilian Amazon expanded markedly between 2008 and 2014, with patches below the 6.25 ha threshold

considered by PRODES increasing over time; those accounted for over 30% of total forest loss in the 2001–2014 period.

2 For a comprehensive review of the international negotiations within the UNFCCC, see, for instance, Pereira and Viola (2020a). On the CBD, see, for instance, Harrop and Pritchard (2011), Henne and Fakir (1999), Oberthür and Rosendal (2014) and Prip (2018).

3 Our analysis starts in 2005 as this is the year that marked the beginning of a sharp reduction in deforestation levels in the Amazon.

4 "Systematic conservation planning is performed to design cost-effective strategies to preserve a subset of the regional biodiversity, including threatened and highly endemic species, unique habitats, special landscapes features, ecosystem processes, and services" (Fonseca & Venticinque, 2018, p. 61).

5 Decree 6.040/2007: http://www.planalto.gov.br/ccivil_03/_ato2007-2010/2007/decreto/d6040.htm

6 Decree 6.041/2007: http://www.planalto.gov.br/ccivil_03/_Ato2007-2010/2007/Decreto/D6041.htm

7 Law 12.512/2011: http://www.planalto.gov.br/ccivil_03/_ato2011-2014/2011/lei/l12512.htm

8 Decree 7.747/2012: http://www.planalto.gov.br/ccivil_03/_ato2011-2014/2012/decreto/d7747.htm

9 Provisional Measure 2186-16/2001: https://www.congressonacional.leg.br/materias/medidas-provisorias/-/mpv/48024

10 Law 13.123/2015: http://www.planalto.gov.br/ccivil_03/_Ato2015–2018/2015/Lei/L13123.htm

11 Data from the Brazilian MMA: https://www.mma.gov.br/images/arquivo/80229/CNUC_FEV20%20-%20C_Biopdf

12 Provisional Measure 910/2019: http://www.in.gov.br/en/web/dou/-/medida-provisoria-n-910-de-10-de-dezembro-de-2019-232671090

References

Albuquerque, F.L. (2019). Coalition Making and Norm Shaping in Brazil's Foreign Policy in the Climate Change Regime. *Global Society*, *33*(2), 243–261. doi: 10.1080/13600826.2019.1571482

Alencastro, C. (2014, August 4). Dilma não criou nenhuma nova unidade de conservação na Amazônia. *O* Globo. Retrieved from https://oglobo.globo.com/brasil/dilma-nao-criou-nenhuma-nova-unidade-de-conservacao-na-amazonia-13479261

Alves, R.J.V., Weksler, M., Oliveira, J.A., Buckup, P.A., Pombal Jr., J.P., Santana, H.R.G., … Caramaschi, U. (2018). Brazilian Legislation on Genetic Heritage Harms Biodiversity Convention Goals and Threatens Basic Biology Research and Education. *Anais da Academia Brasileira de Ciências*, *90*(2), 1279–1284. doi:10.1590/0001-3765201820180460

Barbiéri, L.F. (2019, December 20). Governo Bolsonaro tem aprovação de 29% e reprovação de 38%, diz pesquisa Ibope. *Globo*. Retrieved from https://g1.globo.com/politica/noticia/2019/12/20/governo-bolsonaro-tem-aprovacao-de-29percent-e-reprovacao-de-38percent-diz-pesquisa-ibope.ghtml

Bensusan, N. (2018, November 29). Biologia sintética e pirâmides pedagógicas. *Instituto Socioambiental.* Retrieved from https://www.socioambiental.org/pt-br/blog/blog-do-ppds/biologia-sintetica-e-piramides-pedagogicas

Bittencourt, N.A. (2018). O Brasil e a 14.ª Convenção da Diversidade Biológica: a tragédia anunciada à biodiversidade. *Terra de Direitos.* Retrieved from https://terradedireitos.org.br/acervo/publicacoes/boletins/49/o-brasil-e-a-14-convencao-da-diversidade-biologica-a-tragedia-anunciada-a-biodiversidade/22992

Cabrera Medaglia, J. (2015). Access and Benefit-Sharing: North–South Challenges in Implementing the Convention on Biological Diversity and Its Nagoya Protocol. In S. Alam, S. Atapattu, C. Gonzalez, & J. Razzaque (Eds.), *International Environmental Law and the Global South* (pp. 192–213). Cambridge: Cambridge University Press.

Caetano, M., Mendes, L.H., & Ramos, C.S. (2019, November 7). Governo e agricultores unem forças contra moratória da soja na Amazônia. *Globo.* Retrieved from https://valor.globo.com/agronegocios/noticia/2019/11/07/governo-e-agricultores-unem-forcas-contra-moratoria-da-soja-na-amazonia.ghtml

CAT (2019a), December 2). Brazil: Pledges and Targets. Retrieved from https://climateactiontracker.org/countries/brazil/pledges-and-targets/

CAT (2019b, December 2). Brazil: Current Policy Projections. Retrieved from https://climateactiontracker.org/countries/brazil/current-policy-projections/

CAT (2017). Brazil: Rating. Retrieved from https://climateactiontracker.org/media/documents/2018/4/CAT_2017-11-06_CountryAssessment_Brazil.pdf

Couto, T. (2012). Biodiversity Governance in Developing Countries: Brazil 1990–2010. *Perspectivas – Portuguese Journal of Political Science and International Relations, 9*, 5–29. doi: 10.21814/perspectivas.41

Da Silva Bolzani, V. (2017). Biodiversidade brasileira, regulamentação e o que aprendemos com ela. In H.B. Nader, F. de Oliveira, & B. de B. Mossri (Eds.), *A Ciência e o Poder Legislativo no Brasil: Relatos e Experiências* (pp. 164–173). São Paulo: Sociedade Brasileira para o Progresso da Ciência.

Dallagnol, A., & Rodrigues, C. (2016). Brasil lidera retrocessos na CBD. *Terra de Direitos.* Retrieved from https://terradedireitos.org.br/acervo/publicacoes/boletins/49/brasil-lidera-retrocessos-na-cdb/22415

Dechoum, M.S., Sampaio, A.B., Ziller, S.R., & Zenni, R.D. (2018). Invasive Species and the Global Strategy for Plant Conservation: How Close has Brazil Come to Achieving Target 10? *Rodriguésia, 69*(4), 1567–1576. doi:10.1590/2175-7860201869407

Du Pont, Y.R., & Meinshausen, M. (2018). Warming Assessment of the Bottom-Up Paris Agreement Emissions Pledges. *Nature Communications, 9*, 4810. doi:10.1038/s41467-018-07223-9

Escobar, H. (2014, July 16). Protocolo de Nagoya entrará em vigor sem o Brasil. *Estadão.* Retrieved from https://ciencia.estadao.com.br/blogs/herton-escobar/protocolo-de-nagoya-entrara-em-vigor-sem-o-brasil/

Faleiros, G., & Nascimento, F. (2019, October 1). Sob Bolsonaro, invasões de terras indígenas superam 2018. *Piauí*. Retrieved from https://piaui.folha.uol.com.br/sob-bolsonaro-invasoes-de-terras-indigenas-superam-2018/

FAS (n. d.). Programa Bolsa Floresta. Retrieved from https://fas-amazonas.org/programas/pbf/

Fearnside, P.M. (2018). Challenges for Sustainable Development in Brazilian Amazonia. *Sustainable Development*, *26*(2), 141–149. doi: 10.1002/sd.1725

Federative Republic of Brazil (2015). Intended Nationally Determined Contribution Towards Achieving the Objective of the United Nations Framework Convention on Climate Change. Retrieved from https://www4.unfccc.int/sites/ndcstaging/PublishedDocuments/Brazil%20First/BRAZIL%20iNDC%20english%20FINAL.pdf

Filoche, G. (2013). Domestic Biodiplomacy: Navigating Between Provider and User Categories for Genetic Resources in Brazil and French Guiana. *International Environmental Agreements*, *13*, 177–196. doi:10.1007/s10784-012-9184-z

Folha de S. Paulo (2019, October 10). Minister of the Environment Receives 'Exterminator of the Future Award'. Retrieved from https://www1.folha.uol.com.br/internacional/en/scienceandhealth/2019/10/minister-of-the-environment-receives-exterminator-of-the-future-award.shtml

Fonseca, C.R., & Venticinque, E.M. (2018). Biodiversity Conservation Gaps in Brazil: A Role for Systematic Conservation Planning. *Perspectives in Ecology and Conservation*, *16*, 61–67. doi: 10.1016/j.pecon.2018.03.001

Fuchs, R., Alexander, P., Brown, C., Cossar, F., Henry, R.C., & Rounsevell, M. (2019, March 27). Why the US–China Trade War Spells Disaster for the Amazon. *Nature*. Retrieved from https://www.nature.com/articles/d41586-019-00896-2

Globo Rural (2013, April 24). Deputado alerta para riscos econômicos do Protocolo de Nagoya. Retrieved from http://g1.globo.com/economia/agronegocios/noticia/2013/04/deputado-alerta-para-riscos-economicos-do-protocolo-de-nagoya.html

Gonçalves de Andrade, R.M. (2017). A Biodiversidade Brasileira – Caminhos para Sua Proteção. In H.B. Nader, F. de Oliveira, & B. de B. Mossri (Eds.), *A Ciência e o Poder Legislativo no Brasil: Relatos e Experiências* (pp. 174–182). São Paulo: Sociedade Brasileira para o Progresso da Ciência.

Gortázar, N.G. (2019, May 9). Uma inédita frente de ex-ministros do Meio Ambiente contra o desmonte de Bolsonaro. *El País*. Retrieved from https://brasil.elpais.com/brasil/2019/05/08/politica/1557338026_221578.html

Groen, L. (2019). Explaining European Union Effectiveness (Goal Achievement) in the Convention on Biological Diversity: The Importance of Diplomatic Engagement. *International Environmental Agreements*, *19*, 69–87. doi:10.1007/s10784-018-9424-y

Harrop, S.R., & Pritchard, D.J. (2011). A Hard Instrument Goes Soft: The Implications of the Convention on Biological Diversity's Current

Trajectory. *Global Environmental Change, 21*, 474–480. doi:10.1016/j.gloenvcha.2011.01.014

Henne, G., & Fakir, S. (1999). The Regime Building of the Convention on Biological Diversity on the Road to Nairobi. *Max Planck Yearbook of United Nations Law Online, 3*(1), 315–361. doi:10.1163/187574199X00081

Hochstetler, K.A. (2012). The G-77, BASIC, and Global Climate Governance: A New Era in Multilateral Environmental Negotiations. *Revista Brasileira de Política Internacional, 55* (Special edition), 53–69. doi:10.1590/S0034-73292012000300004

IISD (2010). Summary of the Tenth Conference of the Parties to the Convention on Biological Diversity: 18–29 October 2010, Vol. 9, No. 544. Retrieved from https://enb.iisd.org/vol09/enb09544e.html

IISD (2016). Summary of the UN Biodiversity Conference: 2–17 December 2016. Vol. 9 No. 678. Retrieved from https://enb.iisd.org/download/pdf/enb09678e.pdf.

Imazon (2012). Redução de áreas protegidas para a produção de energia: nota técnica. Retrieved from https://www.amazonia.org.br/wp-content/uploads/2012/05/Nota_tecnica_Tapajos_10mai2012.pdf

ISA (2015a, February 11). Governo atende indústria e ruralistas atropelam votação final de PL de recursos genéticos. Retrieved from https://www.socioambiental.org/pt-br/noticias-socioambientais/governo-atende-industria-e-ruralistas-atropelam-votacao-final-de-pl-de-recursos-geneticos

ISA (2016, June 7). O que o governo Dilma fez (e não fez) pelas Unidades de Conservação? Retrieved from https://www.socioambiental.org/pt-br/noticias-socioambientais/o-que-o-governo-dilma-fez-e-nao-fez-pelas-unidades-de-conservacao

ISA (2015b, May 6). Organizações e movimentos sociais enviam carta à Dilma pedindo veto a projeto de lei da biopirataria. Retrieved from https://www.socioambiental.org/pt-br/noticias-socioambientais/organizacoes-e-movimentos-sociais-enviam-carta-a-dilma-pedindo-veto-a-projeto-de-lei-da-biopirataria

ISA (2014, July 15). Ruralistas bloqueiam ratificação e Brasil passa a ter papel secundário no Protocolo de Nagoya. Retrieved from https://www.socioambiental.org/pt-br/noticias-socioambientais/ruralistas-bloqueiam-ratificacao-e-brasil-passa-a-ter-papel-secundario-no-protocolo-de-nagoya

Kalamandeen, M., Gloor, E., Mitchard, E., Quincey, D., Ziv, G., Spracken, D….. Galbraith, D. (2018). Pervasive Rise of Small-Scale Deforestation in Amazonia. *Scientific Reports, 8*, 1600. doi:10.1038/s41598-018-19358-2

Kröger, M. (2017). Inter-Sectoral Determinants of Forest Policy: The Power of Deforesting Actors in Post-2012 Brazil. *Forest Policy and Economics, 77*, 24–32.

Lima, M., Vale, J.C.E. D., Costa, G. D. M., Santos, R. C. D., Correia Filho, W.L.F., Gois, G.,… da Silva Junior, C.A. (2020). The Forests in the

Indigenous Lands in Brazil in Peril. *Land Use Policy*, *90*, 104258. doi:10.1016/j.landusepol.2019.104258

Lovejoy, T., & Inoue, C.Y.A. (2012). O cluster de biodiversidade. In F. Gaetani, V. Fazio, G. Batmanian, & B. Brakaratz (Eds.), *O Brasil na agenda internacional para o desenvolvimento sustentável: um olhar externo sobre os desafios e oportunidades nas negociações de clima, biodiversidade e substâncias químicas* (pp. 13–71). Brasília: Ministério do Meio Ambiente.

MAAP (2019, September 23). MAAP #110: Major Finding – Many Brazilian Amazon Fires Follow 2019 Deforestation. Retrieved from https://maaproject.org/2019/amazon-fires-deforestation/

Magalhães, M. (2013). Recursos genéticos destinados à produção de alimentos poderão ter regras próprias. *Senado*. Retrieved from https://www12.senado.leg.br/noticias/materias/2013/02/19/recursos-geneticos-destinados-a-producao-de-alimentos-poderao-ter-regras-proprias

Magnusson, W.E., Grelle, C.E.V., Marques, M.C.M., Rocha, C.F.D., Dias, B., Fontana, C.S., … Fernandes, G.W. (2018). Effects of Brazil's Political Crisis on the Science Needed for Biodiversity Conservation. *Frontiers in Ecology and Evolution*, *6*(163). doi:10.3389/fevo.2018.00163

Martins, E., Martinelli, G., & Loyola, R. (2018) Brazilian Efforts Towards Achieving a Comprehensive Extinction Risk Assessment for Its Known Flora. *Rodriguésia*, *69*(4), 1529–1537.

Medeiros, F.L.F., & Albuquerque, L. (2015). A quem pretence a biodiversidade? Um olhar acerca do marco regulatório brasileiro. *Veredas do Direito*, *12*(23), 195–216. doi: 10.18623/rvd.v12i23.533

Ministério do Meio Ambiente (2006a). Deliberação CONABIO n.º 40, de 07 de fevereiro de 2006. Retrieved from https://www.mma.gov.br/estruturas/conabio/_arquivos/Delib_040.pdf

Ministério do Meio Ambiente (2006b). Resolução CONABIO n.º 03, de 21 de dezembro de 2006. Retrieved from https://www.mma.gov.br/estruturas/conabio/_arquivos/resolucaoconabio03_15.pdf

Ministério do Meio Ambiente (2013). Resolução CONABIO n.º 06, de 03 de setembro de 2013. Retrieved from https://www.icmbio.gov.br/portal/images/stories/docs-plano-de-acao/00-saiba-mais/02_-_RESOLU%C3%87%C3%83O_CONABIO_N%C2%BA_06_DE_03_DE_SET_DE_2013.pdf

Ministério Público Federal (2020). Nota técnica nº 1 1/2020/PFDC/MPF, de 3 de fevereiro de 2020. Retrieved from http://www.mpf.mp.br/pfdc/manifestacoes-pfdc/nota-tecnica-1–2020

Mittermeier, R., Baião, P.C., Barrera, L., Buppert, T., McCullough, J., Langrand, O., … Scarano, F.R. (2010). O protagonismo do Brasil no histórico acordo global de proteção à biodiversidade. *Natureza e Conservação*, *8*(2), 197–200. doi:10.4322/natcon.00802017

Moreira, E.C.P., Porro, N.M., & Silva, L.A.L. (Eds.). (2017). *A "nova" lei n.º 13.123/2015 no velho marco legal da biodiversidade: entre retrocessos e violações de direitos socioambientais*. São Paulo: Inst. O direito por um Planeta Verde.

Moreira, R.Z. (2016). *Congresso e Política Externa: A Influência do Legislativo Brasileiro na Tramitação do Protocolo de Nagoya à Convenção da Diversidade Biológica* (Unpublished master's dissertation). Florianópolis, Brazil: Universidade Federal de Santa Catarina.

Muzaka, V., & Serrano, O.R. (2019). Teaming Up? China, India and Brazil and the Issue of Benefit-Sharing from Genetic Resource Use. *New Political Economy*, *25*(5), 734–754. doi: 10.1080/13563467.2019.1584169

Nepstad, D., Schwartzman, S., Bamberger, B., Santilli, M., Ray, D., Schlesinger, P., … Rolla, A. (2006). Inhibition of Amazon Deforestation and Fire by Parks and Indigenous Lands. *Conservation Biology*, *20*(1), 65–73. doi:10.1111/j.1523-1739.2006.00351.x

Newell, P., & Taylor, O. (2020). Fiddling While the Planet Burns? COP25 in Perspective. *Globalizations*, *17*(4), 580–592. doi:10.1080/14747731.2020.1726127.

Nijar, G.S. (2011). *The Nagoya Protocol on Access and Benefit Sharing of Genetic Resources: An Analysis*. Kuala Lumpur: Centre of Excellence for Biodiversity Law.

Nijar, G.S., & Gan, P. (2012). *The Nagoya ABS Protocol: A Record of the Negotiations*. Kuala Lumpur: Centre of Excellence for Biodiversity Law.

Nobrega, C. (2019). COP25: Brazil's Official Presence Diverges Widely From Its Public Persona. *Mongabay Latam*. Retrieved from https://news.mongabay.com/2019/12/cop25-brazils-official-presence-diverges-widely-from-its-public-persona/

Novaes, R.L.M., & de França Souza, R. (2013). Legalizing Environmental Exploitation in Brazil: The Retreat of Public Policies for Biodiversity Protection. *Tropical Conservation Science*, *6*(4), 477–483. doi:10.1177/194008291300600402

Oberthür, S., & Rosendal, G.K. (Eds., 2014). *Global Governance of Genetic Resources Access and Benefit Sharing after the Nagoya Protocol*. New York: Routledge.

Oliveira, E. (2019, December 14). Número de multas aplicadas pelo Ibama em 2019 é o menor em 15 anos, diz relatório do Observatório do Clima. *Globo*. Retrieved from https://g1.globo.com/natureza/noticia/2019/12/14/numero-de-multas-aplicadas-pelo-ibama-em-2019-e-o-menor-em-15-anos-diz-observatorio-do-clima.ghtml.

Orsini, A., & Diallo, R.N. (2015). Emerging Countries and the Convention on Biological Diversity. In D. Lesage & T. Van de Graaf (Eds.), *Rising Powers and Multilateral Institutions* (pp. 258–279). London: Palgrave MacMillan.

Pacheco, A.A., Neves, A.C.O., & Fernandes, G.W. (2018). Uneven Conservation Efforts Compromise Brazil to Meet the Target 11 of Convention on Biological Diversity. *Perspectives in Ecology and Conservation*, *16*, 43–48. doi 10.1016/j.pecon.2017.12.001

Padgurschi, M. C.G., & Joly, C.A. (2017). Brief History of the Brazilian Platform on Biodiversity and Ecosystem Services/BPBES. *Biota Neotropica*, *17*(1), e20170101. doi:10.1590/11676-0611-bn-2017-00010001

Pereira, J.C. (2019). Reducing Catastrophic Climate Risk by Revolutionizing the Amazon: Novel Pathways for Brazilian Diplomacy. In T. Sequeira & L. Reis (Eds.), *Climate Change and Global Development: Market, Global Players and Empirical Evidence* (pp. 189–218). Cham: Springer.

Pereira, J.C., & Viola, E. (2019). Catastrophic Climate Risk and Brazilian Amazonian Politics and Policies: A New Research Agenda. *Global Environmental Politics*, *19*(2), 93–103. doi:10.1162/glep_a_00499.

Pereira, J.C., & Viola, E. (forthcoming). Brazilian Climate Policy (1992–2019): An Exercise in Strategic Diplomatic Failure. *Contemporary Politics.*

Pereira, J.C. & Viola, E. (2020a). Climate Multilateralism Within the United Nations Framework Convention on Climate Change. In *Oxford Research Encyclopedia of Climate Science.* Oxford University Press. doi:10.1093/acrefore/9780190228620.013.639.

Pereira, J.C., & Viola, E. (2020b). Close to a Tipping Point? The Amazon and the Challenge of Sustainable Development Under Growing Climate Pressures. *Journal of Latin American Studies*, *52*(3), 467–494. doi:10.1017/S0022216X20000577.

Prates, A.P.L., & Irving, M. A. (2015). Conservação da biodiversidade e políticas públicas para as áreas protegidas no Brasil: desafios e tendências da origem da CDB às metas de Aichi. *Revista Brasileira de Políticas Públicas*, *5*(1), pp. 28–57). doi:10.5102/rbpp.v5i1.3014

Prip, C. (2018). The Convention on Biological Diversity as a Legal Framework for Safeguarding Ecosystem Services. *Ecosystem Services*, *29*(Part B), 199–204. doi:10.1016/j.ecoser.2017.02.015.

Scarano, F.R., & Silva, J.M.C. (2018). Production and International Trade: Challenges for Achieving Targets 6 and 11 of the Global Strategy for Plant Conservation in Brazil. *Rodriguésia*, *69*(4), 1577–1585. doi:10.1590/2175-7860201869408.

Secretariat of the Convention on Biological Diversity (2014). *Global Biodiversity Outlook 4: A Mid-Term Assessment of Progress Towards the Implementation of the Strategic Plan for Biodiversity 2011–2020.* Montréal: Secretariat of the Convention on Biological Diversity.

SEEG (2019). Emissões Totais. Retrieved from http://plataforma.seeg.eco.br/total_emission.

The Brazilian Report (2019, December 12). How Did Brazil Fare in the United Nations' COP25? Retrieved from https://brazilian.report/environment/2019/12/12/brazil-ricardo-salles-united-nations-cop-25/.

Tigre, M.A. (2017). *Regional Cooperation in Amazonia: A Comparative Environmental Law Analysis.* Leiden and Boston: Brill Nijhoff.

UNEP-WCMC (2014). *Assessing the Biodiversity Impacts of Policies Related to REDD+: Key Considerations for Mapping and Land Use Change Modelling, Illustrative Examples from Brazil.* Cambridge: United Nations Environment Programme World Conservation Monitoring Centre.

Vieira, R.R.S., Pressey, R.L., & Loyola, R. (2019). The Residual Nature of Protected Areas in Brazil. *Biological Conservation*, *233*, 152–161. doi:10.1016/j.biocon.2019.02.010.

Viola, E., & Franchini, M. (2018). *Brazil and Climate Change: Beyond the Amazon*. New York: Routledge.

Wallbott, L., Wolff, F., & Pożarowska, J. (2014). The Negotiations of the Nagoya Protocol: Issues, Coalitions and Process. In S. Oberthür & G.K. Rosendal (Eds.), *Global Governance of Genetic Resources: Access and Benefit Sharing after the Nagoya Protocol* (pp. 33–59). New York: Routledge.

WWF-Brasil (2019, September 9). Queimadas ameaçam espécies em risco na Amazônia. Retrieved from https://www.wwf.org.br/informacoes/noticias_meio_ambiente_e_natureza/?72803/Queimadas-ameacam-especies-em-risco-na-Amazonia.

3 Peru: Hostage to the "Master" of Economy and Finance and Lost in Fragmentation

Peru, the world's sixth most biodiversity country, accounts for over 11% of the Amazon. The lack of infrastructure has hampered economic activity in many areas of the region, helping to keep deforestation levels relatively low. However, deforestation seems to have been continuously increasing since the beginning of the century (Appendix A, Figure A.2, *Geobosques* and Global Forest Watch (GFW) data), putting at risk the forest's ecosystems. This is also the perception of the majority of the interviewees, who provided, in general, the same explanations for such an increase. New data by RAISG suggests that deforestation levels remained virtually stable from 2001 to 2018 – nevertheless, it should be noted that total deforestation for the period 2001–2018 presented by both RAISG in 2020 and the Peruvian government is the same (approximately 23,000 km^2) (Appendix A, Figure A.2).

The official narrative on Amazonian deforestation in the country points to small-scale agriculture by poor, landless migrants, particularly from the Andes region but also from the Peruvian coast, who deforest patches of land smaller than 0.05 km^2 to establish grasslands, subsistence crops and/or commercial crops as well as coca crops for illicit trafficking, as the main driver of deforestation. Once the soil becomes infertile, peasants move to other areas of the forest and deforest new patches. This process is believed to account for approximately 80% of total forest clearing in the Amazon (Ráez Luna, 2019). This narrative was reproduced by several interviewees; however, others raised doubts on its accuracy and/or assumed a more nuanced stance regarding the topic. In fact, as observed by Ravikumar et al. (2016, p. 170–172), the prevailing narrative on Amazonian deforestation

> is based on remote sensing of deforestation patch sizes but not on field data. (…) small patches of deforested land may indicate any

> number of processes, including sustainable fallow management and agroforestry. Moreover, the data underlying the narrative tell us little about the actors driving these processes and their motivations (…)[or] how political negotiations and policies influenced them. (…) [Additionally,] the data-source (…) reports on the *frequency* of deforestation patch sizes and not total *area* deforested.

As highlighted by the authors (and by several interviewees), such a narrative is problematic for a number of reasons beyond the fragile evidence supporting it – from the fact that it puts the blame for deforestation on migrants to the Amazon while apparently overlooking or even excusing other actors as well as disregarding the politics and policies that lie behind deforestation, to the fact that it obscures the development of appropriate responses to the problem (Ravikumar et al., 2016). The contribution of small-scale agriculture to Amazonian deforestation may in reality be much smaller than assumed in the country – a study by De Sy et al. (2015) suggests that smallholder cropland accounts for approximately 40% of forest clearings in the region.

Since the beginning of this century, and particularly over the past decade, illegal mining has become a non-negligible driver of deforestation in the region. Between 2010 and 2017, in the Southern Peruvian Amazon, nearly 650 km^2 were lost to gold mining, an amount representing more than double the total area lost to that activity during the previous almost three decades (Caballero Espejo et al., 2018). Additionally, from the 2000s onwards, areas of the forest have also been cleared for agro-industrial projects, namely palm oil and cocoa (MAAP, 2018, 2020c; Vijay et al., 2018).

In the following pages, we discuss the political, economic and social factors that might explain the potential rising levels of deforestation in the Peruvian Amazon between 2006 and 2019, thus covering the administrations of Alan García (2006–2011), Ollanta Humala (2011–2016) and Pedro Pablo Kuczynski (2016–2018) as well as the first 21 months of Martín Vizcarra's administration (2018–2019). The discussion is divided into four distinct periods: 2006–2011, 2011–2016, 2016–2018 and 2018–2019. The first period is characterized by the government's attempt to open the Amazon to predatory private investment and local resistance to efforts in that direction. It is also, and quite paradoxically, marked by the creation of the Ministry of Environment (MINAM) and the appointment of a prestigious ecologist to head the new institution. However, institutional fragmentation in forest and land management prevented the new ministry from having a significant impact on the governance of the Amazon. Moreover, land concentration deepened, family farmers continued to be marginalized and incentives for commercial and agro-industrial crops contributed to the expansion of the agricultural frontier. During the second period, the country's environmental institutionality consolidated over the first years, but fragmentation continued to hamper the role that MINAM could have in controlling deforestation. Additionally, by the end of this period, the political strength of anti-environmentalist forces became evident, and there were substantial environmental setbacks. Furthermore, the structural causes of deforestation remained unaddressed. The third period is marked by the virtual invisibility of MINAM and political instability, which favored deforestation processes and diverted political and public attention away from environmental issues. The last period is characterized by attempts to fight corruption and crime in the country, which led to some positive advances, but also by a political crisis between the executive and the Congress. Deforestation levels seem to have remained high.

Before starting the analysis, it is important to note that, to understand contemporary Peru and the social, political and institutional environment that underlies deforestation processes in the country, one must recognize that the Peruvian national context continues to be profoundly marked by the effects of, and the trauma caused by, the severe debt crisis lived in the 1980s, the violent insurgency of the Shining Path guerrillas of the 1980s and 1990s, and the policies and institutional arrangements of the authoritarian government of Alberto Fujimori (1990–2000). The 1980s crisis led to the deepening of a massive informal sector in the country (currently, nearly 70% of Peruvians work in the informal sector, and in rural areas approximately 95% of jobs are informal) that weakened civil society groups. These were further debilitated by the Shining Path's murdering of leftist leaders and discrediting of leftist ideologies, and the political, economic, social and repressive measures taken by Fujimori to reduce the size of the state, eliminate whatever social disturbance that could obstruct the functioning of the markets and promote an ideology against collectively organized economic and social processes, in the name of the economic recovery of the country and the fight against the insurgency. Syndicates and native and peasant communities – the latter stigmatized due to the civil war (1980–2000) in the countryside, being associated with violence and terrorism – lost their organizational and influence capacity. At the same time, the power of economic elites rose substantially; a highly expert technocratic elite with power centralized in the conservative Ministry of Economy and Finance (MEF), closely linked to big business, emerged during Fujimori's regime. MEF – which has no solid knowledge or understanding of, or interest in, environmental issues – was and continues to be the dominant actor in planning and decision-making processes in Peru, often displaying autonomy from the country's presidents. For most of the 2000s and the beginning of the 2010s, the strength of economic elites, particularly in the mining sector, consolidated due to the export boom associated with the commodity rising price cycle – ores account for approximately half of Peru's exports. Over the past two decades, the high performance of the economy was somehow decoupled from politics; all Peruvian presidents of this century finished their mandate with extremely low approval rates. MEF's power remains largely uncontested, not only as a result of the weakness of the Peruvian civil society and political opposition – with a weak and fragmented political party system – but also due to a "culture of fear" that has persisted in the country since the Fujimori era, according to which any questioning of the dominant order or any change could lead Peru back to chaos, terrorism, insecurity and the country's exclusion from the financial international system. Moreover, state institutions other than those designed to manage the economy are weak or in some cases lacking – the state's capacity to enforce law and order is thus limited (Interview 20; see, for instance, Crabtree & Durand, 2017; Lust, 2019).

3.1 2006–2011: "The Dog in the Manger", Bloody Resistance and Fragmentation

For President Alan García, whose campaign was greatly financed by the Peruvian banking and mining community (Durand, 2010), forest conservation issues, and environmental issues in general, were far from being a concern – or rather, such issues were a (superficial) concern only in as much as they could hamper Peru's foreign economic and financial relations. The President's plan for the Amazon was to open the region to predatory economic exploitation. In 2007, García published in the daily *El Comercio* a short essay entitled "The Dog in the Manger Syndrome" ("*El síndrome del perro del hortelano*"), where he responded to those who criticized him for promoting extractivism, stating that Peru was not benefiting from the Amazon's natural wealth, because the region was occupied by indolent and poorly educated families, peasants and native communities – seen as second-class citizens – with no capacity to take advantage of the forest's resources and who denied the state the possibility of doing so. The

Amazon's "dogs in the manger" must therefore be replaced by big companies with the means to make the region's resources profitable. The president's metaphor associated the rejection of extractivism with backwardness and underdevelopment – an idea that resonated very well with the conservative business sector. During this period, the number of new hydrocarbon exploration contracts spiked, which resulted in "more Amazonian land area covered by concessions than that at any other time in Peru's history" (Finer & Orta-Martínez, 2010, p. 8); mining concessions on a national level doubled (Observatorio de Conflictos Mineros en el Perú, 2010); commercial and agro-industrial crops expanded (Marquardt, Pain, Bartholdson, & Rengifo, 2019; Vijay et al., 2018); and efforts to pave the Interoceanic Highway connecting the Amazon to the Pacific not only accelerated, but were finalized (Sánchez-Cuervo et al., 2020).

Interestingly though, García's administration was the one that created MINAM in 2008, appointing Antonio Brack, a renowned ecologist, as head of the new institution, plus a number of governmental bodies to guide environmental management: the Environmental Assessment and Enforcement Agency (OEFA), which is an independent oversight body, and the National Service of Natural Protected Areas (SERNANP), both attached to MINAM, as well as the Agency for the Supervision of Forest Resources and Wildlife (OSINFOR), which is an independent agency for overseeing forestry, attached to the presidency of the Council of Ministers (PCM). This was obviously not the result of genuine government recognition of the importance and need for developing and strengthening the country's environmental institutionality – "the dog in manger" philosophy of the President leaving no doubt about it – but a response to foreign concerns over the country's capacity to comply with solid environmental standards and provisions in the context of the approval of both the loan by the Inter-American Development Bank for the second stage of the Camisea natural gas project and the Trade Promotion Agreement (TPA) with the United States, which contains an annex on forest sector governance aimed at promoting biodiversity protection and reducing deforestation in Peru, as well as prospects of negotiating a bilateral trade agreement with the European Union (Interviews 1, 5, 18 and 19).

It should thus come as no surprise that the implementation of the TPA with the United States triggered not only institutional developments in the Peruvian forest governance sector, but the issuing of a massive number of decrees by García under the pretext of implementing the agreement's provisions, many of them which were not required by the agreement and among which were some that limited the protection of the Amazon's lands as forestry areas and undermined communal tenure of land to promote extractivist projects; decrees which, without any public discussion or consultation with indigenous peoples, were converted into the country's new forest law. Amazonian indigenous communities and other regional groups resisted, and the protests, which were associated with terrorism by the president, culminated in armed conflicts, with more than 30 casualties and about 170 people wounded in the Amazonian province of Bagua in June 2009 – a bloody episode known as the "*Baguazo*", which profoundly damaged the image of the government. Pressured by national and international actors, García eventually revoked the most controversial decrees (Interview 4; Merino, 2019; Peinhardt, Kim, & Pavon-Harr, 2019). Against this background, in 2010, the Ministry of Culture was created and the National Institute for the Development of Andean, Amazonian and Afro-Peruvian Peoples (INDEPA) attached to the Vice-Ministry of Intercultural Affairs; a number of policies concerning the rights of indigenous peoples emerged (La Revista Agraria, 2011a; Merino, 2019); and the Congress passed a law on prior consultation with indigenous peoples (Servindi, 2010), which, nonetheless, García eventually did not sign – it was never the government's intent to approve the law, but to make it look as if it was working on the matter (Interview 13). A new forest law was also approved in the final days of García's administration, after a winding, complex process, during which there were unprecedented levels of social participation to compensate for the unilateral and unexpected changes to forestry legislation made by García in the framework of the TPA (Interview 15).

It was within this context that the Minister of Environment, Antonio Brack, taking advantage of both strong indigenous pressure and the fact that the government was seeking to avoid further conflict, was able to obtain MEF's support to create the publicly funded National Forest Conservation Program for the Mitigation of Climate Change (PNCBMCC) in 2010, aimed at conserving 540,000 km^2 of forests (Interview 2) – a proposal which was presented by the Peruvian delegation during the UNFCCC's COP14 in 2008, alongside a pledge to achieve net zero deforestation by 2021. However, important portions of forest cover, many of them with high ecological value and within which a significant proportion of total deforestation occurred, were excluded from the program's mandate. Additionally, the PNCBMCC was endowed with only one effective instrument, which was conditional direct transfers to work with native communities, within whose territories only a small proportion of deforestation takes place (Interview 6; MINAM, 2016).

In fact, MINAM had, and would continue to have, a very limited capacity to address Amazonian deforestation. The forest and land use sectors are exceedingly fragmented in Peru. Institutional fragmentation was pinpointed by most interviewees as a critical variable in hampering the country's capacity to control deforestation, as there is both an overlap and dispersal of functions that complicates governance and control and supervising actions, increases management costs and creates inertia and a sense of de-responsibility among operators. Lack of clarity in rules and procedures creates conditions that allow illegal activities that severely damage the Amazon to mushroom.

With the foundation of MINAM, SERNANP was created and authority over protected areas was transferred from the Ministry of Agriculture (MINAGRI) to MINAM, but despite attempts otherwise, the forest sector was kept within MINAGRI, in the General Directorate for Forestry and Wildlife (DGFFS), which would become extinct in 2013 with the creation of the National Forest and Wildlife Service (SERFOR) – attached to MINAGRI – in accordance with the new forest law. The basic fact of forests being under the responsibility of MINAGRI, which has a productive rather than conservationist orientation, is problematic by itself (although SERFOR is a relatively independent organism, being attached to MINAGRI it has mostly acted according to a productive orientation, focusing on logging forest use rather than on the promotion of more sustainable forest uses (Interviews 9 and 14). Why was the forest sector kept within MINAGRI? First, forests are mainly seen in the country from the perspective of their utility function as sources of wood, and it would become clear in the future that MEF had plans for transforming the productive potential of Peruvian forests into a driver of the country's economic growth (Interviews 6, 16 and 19; MEF, 2019). Second, given that Peru is a weakly industrialized country and heavily dependent on commodities, the idea of transferring control of any productive functions to MINAM could obviously only encounter resistance – particularly in light of the fact that "the Peruvian Amazon is a chaotic combination of subsistence economy with informal economy, illegal economy and a formal economy based on the extraction of natural resources, [logging included,] which are in many cases intertwined" (Interview 16), as shall be seen. Third, the logging industry, in spite of its still modest contribution to the country's economy (logging represents less than 1% of the GDP), is a powerful one; illegal logging is, alongside drug trafficking and illegal mining, at the top of the list of criminal activities that mobilize the greatest amounts of money in Peru (De Echave, 2016). The logging industry exerts much influence at regional/local level and since the mid-2000s, the strength of regional political parties has been growing (Mcnulty, 2017); moreover, the weakness and fragmentation of the Peruvian party system, with a relatively individualized Congress and inexperienced politicians, facilitates lobbying (Mujica, 2014). We will return to these topics later in this chapter. However, it is thus unsurprising that another attempt at transferring the forest sector to MINAM was rejected in the context of the approval of the new forest law, with many congressmen opposing the idea (Interview 15).

Nevertheless, because biodiversity and ecosystem management were the responsibility of MINAM, and with the creation of the PNCBMCC, their officials have always considered that the institution was also entitled to intervene in forest governance. The ministry's functions and, most importantly, how those functions ought to be articulated with those of other ministries were not clearly defined. This has caused some tension between MINAM and MINAGRI, which have never found a solid way of collaborating and fulfilling the objectives of both sectors (Interviews 5 and 6). To complicate the situation further, decentralization in the forest sector was unblocked, and most executive and monitoring responsibilities on forestry began to be transferred to regional governments (Legislación Ambiental, n. d.). However, as shall be seen, this was a process filled with serious deficiencies which, in addition to the fact that it further fragmented forestry management in the country, severely affected the Amazon, according to most interviewees. Although the decentralization law had been approved during the previous administration, the process was moving forward slowly, and the transfer of forestry responsibilities to the regions was paralyzed; MINAGRI had not yet done so (Interview 6). A ministerial crisis in 2008, precipitated by a corruption scandal involving several political actors behind the government, led García to appoint a regional governor as prime minister to have the support of the regional governments, whose political influence grew as a result (Meléndez, 2009) – within this context, decentralization advanced further (Interview 6). The new responsibilities of regional governments would also be included in the new forest law. Finally, still regarding fragmentation in the forestry sector, one cannot forget the existence of OSINFOR.

Institutional fragmentation is, furthermore, a problem in the agrarian sector regarding land use change issues (Piu Deza & Galván Gildemeister, 2015). With the institutional reforms triggered by the formation of MINAM, the General Directorate of Agrarian Environmental Issues (GDAEI) was also created within MINAGRI, which had the responsibility of approving environmental licensing activities in the agrarian sector, classifying land and authorizing decisions on land use change to develop agroecological zoning, and addressing, in collaboration with MINAM, the climate change issue. Regional governments became in charge of managing and approving land use changes following MINAGRI's binding decision. MINAM, in turn, as the governing body for land use planning in the country, was assigned the task of developing technical ecological-economic zoning studies not only to promote the sustainable use of natural resources, but also to ensure legal certainty for businesses and avoid social conflicts, providing additional land use information to complement the studies conducted by MINAGRI, including data on biodiversity, watershed management, mineral reserves and local populations – such ecological-economic zoning studies should be the basis for developing land use plans. However, for reasons that will be addressed in the next subsection, including the fact that regional governments were given executive responsibilities regarding land use planning, the process would be slow. Consequently, land use change for agricultural purposes would continue to prevail. For classifying land, the GDAEI makes its decisions according to the Best Land Use Capacity to evaluate an area's suitability for agriculture, a tool whose methodology omits the presence of living trees on the land (Interview 15; see also Glinskis & Gutiérrez-Vélez, 2019; Servindi, 2017) – since the GDAEI assumed this function in 2009, permits for land use change in forest areas, particularly for permanent industrial or export crops, increased significantly (Dourojeanni, Ráez Luna, & Valle-Riestra, 2016).

Simultaneously, historically marginalized rural populations continued to lack the necessary technical and financial assistance that could allow them to cultivate land already deforested and develop sustainable family farming. Seen as inefficient, unproductive and primitive, traditional farming systems, which "can in fact be key parts of a sustainable land use agenda" (Ravikumar et al., 2016, p. 175), have been depreciated by successive governments – openly by García's administration. Governments have, alongside the private sector, concentrated resources on large agro-industrial projects, particularly on the coast, while the highlands and the jungle remained marginalized, and promoted agricultural intensification and commercialization, with credit incentives for monoculture cash crops; these have also been boosted by alternative development programs designed to help coca-growing families to abandon illicit crops. Consequently, new settlers have arrived in the Amazon region attracted by the economic opportunities associated with commercial and agro-industrial crops, and land has begun to be rented for monocultures, increasing land scarcity and pushing the agriculture frontier further into the landscape (see, for instance, Bennett, Ravikumar, & Paltán 2018; La Revista Agraria, 2011a; 2011b; Marquardt et al., 2019; Ráez Luna, 2019; Ravikumar et al., 2016). This trend would continue until the present. As explained by one interviewee (Interview 17), governmental policies

> have been disrupting the functioning modes of local communities. Agricultural production that used to be mainly for domestic self-consumption or communal exchange has increasingly been transformed. Producers have begun to target the market. This has led to changes in production patterns, which in turn have also led to changes in cultural and social patterns at the local level.

In addition to the agro-industrial push, the corporativist push of García's administration accelerated land concentration in the country (La Revista Agraria, 2009; Ráez Luna, 2020). This translated into additional processes of migration to the Amazon. Mining concessions, for example, are a highly influential factor in forcing people to migrate from the highlands (Interview 7; Cooperacción, 2014).

How did/do settlers arrive in the region? The Interoceanic Highway and the secondary road network which emerged as a result, for example, opened the door to many migrants to the Amazon and became major drivers of deforestation (Sánchez-Cuervo et al., 2020). Moreover, roads have also allowed the gradual integration of parts of the region into the country's market economy, incentivizing the expansion of the agricultural frontier (Marquardt et al., 2019). Additionally, throughout the 2000s (and the whole period addressed in this chapter), illegal logging proliferated, with logging concessions being widely used to launder timber illegally extracted from outside concession areas (Finer, Jenkins, Blue Sky, & Pine 2014) – illegal logging is not a significant direct driver of deforestation in the Amazon, but an important indirect one, as forest extraction paths/roads are also often the access routes used by migrants to settle in the region (Interviews 4, 8, 11 and 15). Roads and pipeline routes connecting to hydrocarbon drilling platforms have also converted into indirect drivers of deforestation, as they allow access to previously remote areas (Interview 14).

The effects of development and infrastructure projects, either concluded or planned, in increasing land prices and boosting land trafficking, ought also to be considered. One interviewee (Interview 3) asked, "How can we know for sure whether the main cause of deforestation is not land speculation? With the data we have currently available, we cannot know. Note that land clearing and economic exploitation of the area are required to establish property claims in Peru". In fact, studying deforestation processes in the Amazonian regions of Amazonas and San Martín in the period 2007 to 2015, Shanee and Shanee (2016) found that land trafficking can be a highly lucrative activity and that migration is frequently organized and led by land traffickers. Forest clearing for money laundering purposes, given the rapid expansion of illegal mining, for example, also needs further scrutiny in this context (Interview 3; Hill, 2016).

Regarding illegal mining, in 2010, pressured by Antonio Brack, the government initiated an offensive against that activity in the Amazonian region of Madre de Dios. An "urgent decree" was issued outlawing the dredges operating on the rivers, limiting mining to an area of less than 5,000 km^2 in the valley of the river Inambari, where the majority of concessions granted in the past existed, and suspending acceptance of new mining claims for a year. Once the prohibition expired, nevertheless, miners lined up at the regional mining office to stake new claims and soon a new mining hotspot would emerge (Fraser, 2011). Additionally, the state's security forces were sent to the region to remove miners and destroy their equipment, but these actions had limited success. In many cases, small miners work on behalf of larger mining enterprises (Caballero Espejo et al., 2018). Consequently, equipment destroyed was easily replaced. Furthermore, authorities were outnumbered by miners (Fraser, 2011). Negotiations to formalize the activity of small miners were also unsuccessful, partly due to the aggressive approach to the problem by the government, which raised tensions (Interviews 1 and 10; SPDA Actualidad Ambiental, 2011). Moreover, and most importantly, as shall be seen later in this chapter, the state at all levels is infiltrated by illegal mining interests, and there is little political will to solve the problem, which would require socioeconomic reforms and incentives coupled with law enforcement (Interviews 1 and 8). Since 2003, as a result of rising metal prices, miners' organizations have consolidated and the economic, political and social influence of the informal and illegal mining sectors increased notably (De Echave, 2016). Last but not least, institutional control over artisanal and small-scale mining was also fragmented during this period, with regional governments having been responsible for managing the sector since 2008 (Dourojeanni et al., 2016).

On a positive note, the creation of protected areas continued during this period. García's administration was the one during which the largest number of protected areas in the country's history was created – this was clearly the result of the work of Antonio Brack, financially supported by the German government. In terms of extension, nevertheless, García's administration only ranks fourth, with the protection of approximately 18,200 km^2 of forest – way below the previous administration, of Alejandro Toledo (2001–2006), during which an extension of almost 47,000 km^2 of forest was protected (Dourojeanni, 2017). Ever since the Bagua incident, Amazonian indigenous peoples have constructed barriers to MINAM's policy of protected areas, pressuring for the titling of their territories to be considered more equally alongside the creation of those areas. The historically low confidence in the state by indigenous communities deepened; they became suspicious of governmental initiatives in the region, fearing that their territories could be taken from them and given to private companies or non-governmental organizations (NGOs) (Interviews 4, 14 and 15); moreover, opposition from economic groups to protected areas would also grow in the following years. García's administration is, however – excluding the administration of Valentín Paniagua (2000–2001), who was in office less than a year – the one which demarcated and titled the least extension (nearly 1,500 km^2) of indigenous territory in the country's history.[1]

In sum, during García's administration, political options driving deforestation since the 1990s, such as the marginalization of family agriculture, incentives for commercial and agro-industrial crops, and the promotion of environmentally-damaging development and infrastructure projects, remained and, in many cases, deepened. Additionally, the seed of another key political driver of deforestation was planted – institutional fragmentation in critical sectors affecting the Amazon.

3.2 2011–2016: More Fragmentation and Environmental Ups and Downs

In July 2011, Ollanta Humala, a leftist candidate promising a "great transformation", won the presidential election. The new president appointed to his cabinet several representatives of the groups who supported his election, but governmental economic positions from previous administrations did not suffer significant changes – Peru's National Confederation of Private Business Institutions (CONFIEP) initiated contact with Humala following the first round of elections in May, pushing him to a more moderate position (Durand, 2016).

Ricardo Giesecke, a physicist and expert on environmental issues, became Minister of Environment, supported by Hugo Cabieses and José De Echave as vice-ministers – three figures who, from the perspective of the business sector, could be considered as radicals (Interview 7). Their passage through the ministry was very short. With a highly ambitious agenda including proposals such as (a) the transferring of the forest sector and capacities of environmental impact assessment analyses to MINAM (note that those were dispersed under the authority of several ministries, meaning that each sector was in charge of approving environmental impact assessment studies for projects under its domain); (b) the implementation of a state policy against informal/illegal mining based on creating conditions for rural development and population retention programs to control migration from the highland region to the Amazon; (c) the conversion of the National Center for Strategic Planning (CEPLAN) into a ministry, thus weakening MEF's role on national planning decisions; (d) the approval of a law on land use planning, for which Giesecke's ministry was already developing a proposal; (e) the acceleration of ecological economic zoning, with a legally binding character; and (f) the decentralization of MINAM, with the creation of environmental regional authorities (Interview 1; MINAM, 2011), the new cabinet could only have violently shocked with established economic interests. The tipping point was the mining conflict in Conga, during which Giesecke publicly raised doubts over the reliability of the associated environmental impact assessment study (Interviews 1 and 12). The protests against the Conga project, for which massive investments had already been made, led to its suspension; after that, CONFIEP and other actors demanded that the government guaranteed protection for their investments, and in late 2011, Humala asked for the resignation of several ministers and advisers (Durand, 2016), Giesecke and his cabinet included.

Within this context, decentralization in the forest sector continued to advance, while conditions for decentralization in general fell short. Given that regional governments had been assigned executive and monitoring responsibilities to ensure the sustainable use of natural resources within their jurisdictions and conduct land use planning, the allocation of resources and the creation of solid conditions for decentralization – which could equip regional governments with the means to successfully undertake those tasks – were not in the interest of economic elites, as that could become an obstacle for private investments (Leyva, 2012). Regional governments were assigned critical tasks without being given the necessary resources and capabilities for accomplishing them, and, most importantly, without mechanisms for ensuring transparency and effective oversight – corruption levels at the regional and local levels are great. In the absence of land use planning and in a context marked by corruption, areas of the Amazon are more easily concessioned to the private sector, occupied for informal or illegal activities, classified as agricultural lands, trafficked and opened with new infrastructure. Most interviewees indicated the lack of a law on land use planning and the transfer of functions to regional governments amid institutional fragmentation and lack of clarity in rules and procedures as key variables to understand the last decade's rising levels of deforestation in the Amazon. It is also important to note that, during most of this period, as a result of the transition of the forest sector from the DGFFS to SERFOR and the long process of regulating the forest law approved in 2011 (which was only concluded in 2015, following high levels of social participation, especially by indigenous communities, once more to compensate for the actions of García's administration), the forest sector was left with no clear guidelines for regional governments to execute their functions (Dourojeanni et al., 2016). In brief, to understand deforestation in the Peruvian Amazon, one must look at the articulation of different actors and activities – informal, illegal and formal – in the region, without losing sight of the institutional and political environment that intentionally or unintentionally makes such articulation possible (Interviews 9, 11, 12 and 15).

The logging industry is a powerful one on a local level; bribes to authorities are commonplace not only to ensure conditions for illegal logging and timber laundering, but also to facilitate (through land titles, for example) the construction of the roads needed for extracting timber – the same roads that are used by other actors to access the region. Exploitation, forced labor and subjugation of local vulnerable communities at the hands of the logging industry are also common in the Amazon. Additionally, there is evidence of association between logging, mining and even drug trafficking, which points to the existence of regional/local power networks cemented in different illegal industries, and also explains the power of the logging lobby and regional governments in influencing decision-making in the forest sector. Moreover, the logging industry has been denounced for being infiltrated into regional indigenous land titling agencies and interfering in the titling process to get access to new forest areas to harvest timber and/or sell the land (Interviews 9, 10, 11, 14 and 19; Pachico, 2012; Romero, 2014).

The intertwinement between informal, illegal and legal mining in Peru's gold supply chain, including the processes through which illegal gold from informal mining activities enters the legal market, has also been recently scrutinized. Possibilities to launder gold are facilitated not only by local corruption, but also by a permissive attitude by the state in relation to the activities of gold buyers (Van der Valk, Bisschop, & van

Swaaningen, 2020). Illegal miners have been financing several political forces and increasing their presence in the national Congress and in subnational governments – in 2014, the President of the Federation of Miners of Madre de Dios (FEDEMIN) even became regional governor – thus strengthening lobbying activities against the prosecution of illegal mining, and collaborating with formal metal processing and marketing companies, which launder their illegally extracted metals (De Echave, 2016). Unsurprisingly, MINAM's initiatives during this period to prohibit mining in fragile ecosystems faced harsh opposition from MEF, the Ministry of Energy and Mines (MINEM) and the PCM (Dourojeanni et al., 2016). Furthermore, legal requirements established during this period for miners to make the transition to formality are hardly applicable to alluvial mining in the rainforest, making it almost impossible for those who intend to formalize their activities to do so. Criminalization and repression continued to be the preferred modus operandi, to make it look to environmentalists and the international community as if the government was committed to protecting the Amazon from illegal mining (Bernet Kempers, 2020). Finally, one must not neglect the fact that not only does the logging industry exert pressure for the construction of roads; the informal/illegal mining sector also does so (Dourojeanni, 2015).

Another example of the link between different actors is the buying of small tracts of titled land from farmers by agro-industrial companies as a way of skipping the Peruvian legislation. Land titles can only be given if the land was classified as suitable for agricultural development, but regulations are in practice not enforced for titling small areas of land to settlers, because titling is conceived of as a way to regularize an existing occupation of land. Regional authorities have been exposed, for example, for giving new land possession certificates to settlers operating in alliance with agribusiness companies aimed at establishing large plantations in the Amazon (Dammert, 2018). As for land trafficking, this activity is also facilitated by corruption on a local level or directly promoted by regional and local authorities, with officials providing false papers proving previous ownership of the land (Shanee & Shanee, 2016), the launching of massive titling campaigns to attract votes or give land to family, friends and acquaintances, and the organization of invasions to then grant land property rights (Sierra Praeli, 2018). The dynamics of land trafficking are also significantly influenced by agriculture loans, credits and assistance provided by public and private institutions, with some projects not requiring land titles or a minimum time of occupancy before credits can be granted (Shanee & Shanee, 2016).

During this period, the encroachment of agro-industrial companies in the Amazon was exposed and stopped by local civil society groups in collaboration with journalists, international NGOs and supervisory state authorities, who were able to attract international attention to the increasing contribution of large-scale monocultures (cocoa and palm oil) to forest clearing, and successfully prevented deforestation from expanding (López Tarabochia, 2017) – without such pressure, it would have been relatively easy for regional or local governments to regularize the situation of those companies. This was not a planned alliance or campaign, but rather a coincidence of actions, with groups and organizations joining forces along the way, and journalists who embraced the cause and have continued to follow and report on the case (Interview 19). Since then, the Peruvian Amazon has become a more monitored region, making it harder to establish large cocoa or palm oil plantations (Interviews 4 and 11). Campaigns like this remain, nevertheless, isolated or punctual. Most NGOs are government regulated or depend on joint projects and contracts with public authorities, and governmental control on their work is tight – most NGOs are therefore too soft with the government or avoid parting with the status quo (Interviews 4, 15 and 19). Moreover, it is difficult for NGOs to move within the highly complex Peruvian institutionality (Interview 19). As for indigenous communities, they are, beyond the broad themes that affect them all, such as prior consultation and the titling of their territories, too divided, with internal struggles for power (and recently denunciations for mismanagement of funds by the Inter-Ethnic Association for the Development of the Peruvian Jungle – AIDESEP), which weakens their prestige and capacity to influence decisions, and makes them more vulnerable to manipulation by certain sectors (Interviews 1, 4, 9 and 19; see also Chirif, 2012; AIDESEP, 2019).

It should be noted, however, that in spite of the successful campaign against agro-industrial companies, during this period, Company-Community Partnerships (CCP) for environmental and socially responsible palm oil production between private or state-owned companies and rural communities or villages were formed and a national plan for the sustainable development of palm oil for the period 2016–2025 was launched. CCPs have been seen by companies as a way of accessing land and avoiding conflicts by working directly with local communities, which helps them build a positive public image. Nevertheless, a recent study focusing on the Amazonian region of Ucayali showed that

> CCP farms exhibited much more deforestation then neighboring farms that did not grow oil palm (…)[and that] there may be other

> deforestation linked to displacement of traditional subsistence farms and their colonization of new forest areas. (…) Oil palm is now the third largest agricultural driver of deforestation [in Peru] (Bennett et al., 2018, p. 30; p. 39).

Although Humala's rhetoric was much more nuanced, with the President criticizing "the dog in the manger" philosophy, his government kept the policy orientation of García's administration, with the promotion of large investments by extractive industries, facilitating their access to land, agricultural intensification and the marginalization of family agriculture, in addition to making virtually no advances in titling and regularization of land tenure rights – a task under the responsibility of MINAGRI since 2013 – which would be key, for example, to facilitating investments in already deforested or degraded land and sustainable forest concessions, and curbing land trafficking. This is, to a very significant extent, the result of MEF's orientations, which mobilizes the main resources for the sector and decides on their allocation. MINAGRI's budget has been insufficient to respond to necessities in the sector – whose contribution to Peru's economic growth is much lower than that of mining – and has continued to be channeled into agro-industry on the coast and commercial, export crops (Interviews 1, 5 and 15; Castillo Castañeda, 2016; Eguren, 2016). Moreover, regardless of MEF's control of decision-making, halting deforestation has simply never been a priority for MINAGRI (Interviews 5, 6, 9 and 15; Dourojeanni et al., 2016). Another aspect worth considering is that the new forest law, which is supposed to be a pivotal tool to promote forest conservation, as it includes, for example, agroforestry as an activity that can be formalized, became excessively long and complicated to implement as a result of its highly participative process of regulation (Interviews 6, 8, 10 and 15). Additionally, access to the Amazon by migrants or to new areas within the region by displaced residents continued to be facilitated by the expansion of the road network (RAISG, 2020), at a time when social programs targeting extreme poverty were being implemented (this was the greatest shift in Humala's administration compared to the previous ones) – this might have allowed more people to migrate to the Amazon, with a few more resources to clear the forest (Interviews 7 and 21).

There were, however, some positive advances during this period as well, which would fuel opposition by the country's anti-environmentalist forces. At the beginning of his mandate, Humala, demarcating himself from García and pressured by indigenous communities (Interviews 15 and 22), enacted the prior consultation law; the delimitation and titling of indigenous communities was also given an important impulse during this period (approximately 21,000 km^2), with his administration being overtaken only by that of Alberto Fujimori (nearly 64,000 km^2 over 10 years).[2] Nevertheless, it should be noted that, in Peru, property rights to indigenous communities are granted only for living and agriculture areas, while a contract for cession of use is granted for the forest area, which means that indigenous peoples can use forests fulfilling certain conditions, but cannot dispose of them (Ráez Luna, 2020). In other words, "the process is based on a conception of agricultural use of land, which does not reflect the use that indigenous peoples make of Amazonian land" (Interview 9). Additionally, for most communities, it is too burdensome to watch over their multiple and dispersed boundaries; this makes their territories vulnerable to invasion by loggers and land traffickers (Ráez Luna, 2020).

In the forest sector, OSINFOR, headed by Rolando Navarro, who represented a shift in the government's commitment to fight illegal logging, focused on dismantling logging mafias and seizing illegal timber, and undertook *Operación Amazonas*, a well-succeeded international operation in coordination with SUNAT, which is the country's custom agency, and INTERPOL, during which massive irregularities in timber exports were detected. However, the logging industry retaliated, and Rolando Navarro was eventually removed from office in 2016 (Guidi, 2016). A report issued by the US government that same year found that 90% of the timber exported to the United States was illegally logged (Interagency Committee on Trade in Timber Products from Peru, 2016).

Other important advances came from MINAM. Manuel Pulgar Vidal, who was Director of the Peruvian Environmental Law Society (SPDA), and several other figures coming from the NGO world, assumed the ministry in the beginning of 2012, after Giesecke's exit. The strategy of Minister Pulgar Vidal was to avoid the themes which he knew he could not advance and focus on finding key, timely opportunities that could allow MINAM to make progress (Interviews 4, 5 and 19). For example, during the Rio + 20 environmental conference in 2012, Pulgar Vidal and his team showed President Humala that environmental protection was also a matter of development and made him sensitive to the need to strengthen the country's environmental institutionality and make environmental impact assessment studies more transparent. Within this context, Humala supported the creation of the National Service of Environmental Certification for Sustainable Investments (SENACE), an agency attached to MINAM, which was assigned the task of assessing the environmental viability of the most complex investment projects in the country (Interview 5). OEFA was also strengthened during part of this period, having greatly increased its budget as a result of a new fine regime and a new financing mechanism according to which companies had to contribute up to 1% of their annual turnover to finance public regulatory entities – between 2011 and 2015, the number of direct supervisions by the agency increased fivefold, reaching more than 5,600 per year. Besides, through a communication campaign that attracted the support of the national media, MINAM also managed to broaden OEFA's capabilities in the illegal mining sector and take action against some operators (Dourojeanni et al., 2016; Valle-Riestra, 2017). Moreover, the announcement that Peru would host COP20 in 2014 – Pulgar Vidal made climate change MINAM's flagship issue – opened several doors for attracting international financing for the management of Peru's forests (Interview 5), among them being a 300-million-dollar agreement for verified results with Norway to halt deforestation (MINAM, 2016). Pulgar Vidal and his team were very skilled in obtaining foreign funding for halting deforestation in the Amazon (Interviews 5 and 11) – however, as previously mentioned, MINAM is very limited in its capacity to influence decisions driving forest clearing. Additionally, *Geobosques*, a satellite forest monitoring tool allowing the tracking of forest cover and loss, was also created (Charpentier, 2016). By the end of Humala's administration, a set of climate change-related policies, such as the national forestry and climate change strategy for reducing deforestation, were approved. Lastly, an extension of nearly 17,300 km^2 of forest was protected during this period (Dourojeanni, 2017); however, it should also be noted that aggressions to protected areas increased (Dourojeanni et al., 2016).

Despite some positive advances, the strength of anti-environmentalist forces consolidated from the middle of this period as a result of falling commodity prices, which translated into economic slowdown; Humala's political inexperience and his party's undisciplined, disloyal legislative bench (Vergara & Watanabe, 2019); public political opposition by Alberto Fujimori to the President following the latter's refusal to the former of a presidential pardon in 2013 (Fujimori had been sentenced in 2009 to 25 years in prison for crimes against humanity) (Barrenechea, 2014); the stability, discipline and strengthening of *fujimorismo*, which has ties to extractivist sectors, in the Congress; and the resulting consolidation of legislative opposition (Muñoz & Guibert, 2016). Within this context, a series of legal norms – the most controversial of them having been fully elaborated by MEF – reducing and simplifying environmental procedures, facilitating investments in the mining and hydrocarbon sectors, and weakening MINAM were issued: (a) ecological-economic zoning was determined as non-legally binding and land use planning policy established as the subject of approval by the PCM (while a bill on land use planning remained stagnated in Congress); (b) OEFA was debilitated, with a three-year moratorium on its sanctioning power and a severe reduction in environmental fines, making it more advantageous for companies to pay a fine than to prevent environmental harm; (c) MINAM lost the autonomy it had to establish reserved zones, which is the first step to creating protected areas; (d) SUNAT's supervisory capacity was limited; and (e) deadlines for the approval of environmental impact assessment studies were reduced (see, for instance, Dourojeanni et al., 2016; Salazar, 2016; Servindi, 2017; Valle-Riestra, 2017), to name a few examples.

In sum, during this period the structural causes of deforestation remained unaddressed or were aggravated, with the deepening of institutional fragmentation and the consolidating power of anti-environmentalist forces further cementing the conditions that allow the regional articulation of actors and activities driving many deforestation processes in the Amazon.

3.3 2016–2018: The Environment Off the Radar, (More) Lawlessness and Political Instability

In 2016, Pedro Pablo Kuczynski, a figure who openly represented the interests of Peruvian economic elites, won the presidential election against Alberto Fujimori's daughter, Keiko Fujimori. Nevertheless, as observed by Vergara and Watanabe (2019, p. 36), "his victory did not mean that Peruvians had embraced the status quo. After election officials barred two more appealing outsiders from the race, (…) between neoliberal continuity and the possibility of democratic backsliding, (…) Peruvians chose the former". Furthermore, during the presidential campaign, Kuczynski promised to fight corruption and criticized the corrupt and authoritarian background of *fujimorismo*; nevertheless, for reasons addressed below, this was a period of growing lawlessness in the Amazon.

Under Kuczynski's presidency, MINAM continued to be weakened. The attention that the ministry had attracted with the hosting of COP20 in 2014 was an inconvenience for the business sector. The team chosen by Kuczynski to head MINAM, although formed by competent people with experience in the environmental sector, was ordered to keep a low profile, avoid any confrontations and facilitate (predatory) economic growth, thus aligning with MEF. Moreover, the ministry lost highly qualified technicians and employees. Consequently, MINAM withdrew into itself. The new head of the institution, the economist

Elsa Galarza, had a very moderate discourse, avoiding sensitive issues such as forest conservation, and focused on advancing solid-waste management (Interviews 1, 4, 5, 7, 9, 10, 12 and 15). The PNCBMCC, which had become the focal point of several projects aimed at forest conservation, was fragmented; different projects were put under the responsibility of different technical teams, whose coordination was very limited, which hindered their implementation and blocked previous advances on intersectoral proposals and projects designed to work with regional governments and civil society groups – a problem that has persisted to this day (Interviews 6, 7 and 15). In addition, MINAM relinquished the authority as governing body for national land use planning policy, with its functions being restricted to "environmental land use planning" – a classification of land use planning with no legal definition (Valle-Riestra, 2017) – and reduced environmental standards for mining projects (Programa de Política y Gobernanza Ambiental de la SPDA, 2017).

Other notable aspects with implications for the Amazon during Kuczynski's presidency were the initiation of a forestry executive bureau by MEF, with the participation of logging industrialists who, in some cases, are clearly involved in illegal logging (Interviews 6 and 19), as part of the ministry's aim of activating new engines for national economic growth in the face of the continued downturn in international commodity prices (Andina, 2017), and approval of the construction of a set of new roads along Peru's border with Brazil in the Amazonian regions of Ucayali and Madre de Dios (Erickson-Davis, 2018). Conversely, as a result of pressure by the United States in light of the finding that about 90% of all sourced timber from Peru was illegal, Kuczynski's administration pledged to improve the traceability of timber flows (Charpentier, 2016), but rules requiring loggers to make timber flows traceable would only be approved in 2019 (El Peruano, 2019), under a different administration. Additionally, the environmental crimes of illegal logging and mining were included in the law against organized crime (Ipenza Peralta, 2018).

Kuczynski's administration did not last two years in office. The *fujimorista* party, *Fuerza Popular*, which had performed very well in the congressional elections, ensuring a clear majority, mobilized to destabilize and paralyze Kuczynski's mandate, whose party had won only 18 seats in Congress (McNulty, 2017) – an unparalleled situation in democratic Peru; no other President had had so little congressional support. Moreover, the Brazilian Odebrecht scandal implicated the Peruvian president (and all other previous presidents of this century). Consequently, an impeachment vote took place by the end of 2017.

The President was kept in office with the surprising abstention of some members of *Fuerza Popular* – granting a presidential pardon to Alberto Fujimori, and thus contradicting his campaign promise, Kuczynski obtained the support of Kenji Fujimori (Keiko's younger brother) and ten of his loyal legislators. However, Kuczynski's pardon earned him the opposition of members of his legislative bench and other parties, and a new impeachment procedure began. However, the President resigned in March 2018, before the second impeachment vote took place, following the release of a video showing members of his government trying to buy votes from Kenji Fujimori to avoid Kuczynski's downfall (Arce & Incio, 2018).

Amid political instability and corruption scandals, alongside rising levels of crime and public insecurity, and with MINAM's silence, environmental issues practically disappeared off the political and public radar. Corruption and lawlessness grew at regional and local levels, as the government had no interest and/or capability (or less interest and/or capability than usual) of confronting regional power groups and governments, and the media was absorbed in the covering of corruption scandals and the political crisis – this favoured deforestation processes in the Amazon, with the approval of projects and land concessions, among others (Interviews 11, 13 and 14; Sierra Praeli, 2018); record high levels of gold mining (MAAP, 2019); the invasion of protected areas by drug traffickers (Sierra Praeli, 2018); and the occupation and destruction of forest land by a religious group associated with organized agricultural activity called the Mennonites (MAAP, 2020a). Additionally, it should be noted that governmental weakness led to key positions and institutions such as SUNAT or the People's Advocacy being left in the hands of the legislative power or penetrated by employees with close ties to *Fuerza Popular* (Arce & Incio, 2017) – to which, as previously mentioned, a part of logging and mining interests are connected. It is interesting to observe, though, that investigations triggered by the Odebrecht case led to several regional governors being sentenced to jail for corruption during this and the next period.

3.4 2018–2019: Fighting Crime (?) Amid Escalating Political Crisis

With Kuczynski's resignation, his Vice-President, Martín Vizcarra, who was in Canada as ambassador, assumed the country's presidency. Demarcating himself from the scandals of the previous administration, the president took measures to fight corruption and crime. Taking advantage of *Fuerza Popular*'s internal problems – with Kenji's dissidents and scandals linking the party to corruption and influence-peddling networks leading to Keiko's prison sentence – as well as of public support for his actions against corruption, and displaying detachment from power and a good capacity to manage the executive–legislative relationship throughout 2018, Vizcarra promoted a number of institutional reforms targeting the judiciary system and the Congress via referendum. He also supported the special team of anti-corruption prosecutors created in 2016, which assumed the leadership of all investigations related to the Odebrecht case in 2018. However, the executive's attempt to approve a political reform that, among other measures, included the lifting of parliamentary immunity, culminated in an institutional crisis in September 2019, with the president dissolving Congress and the Congress suspending the president: massive popular approval (85%) of the dissolution of the Congress, high levels of presidential approval (79%) and support by the military and the police led to the crisis being solved in favor of the executive, and congressional elections would be held in January 2020. In short, Vizcarra became an "institutionalist *caudillo*" (see Ponce de León & García Ayala, 2019 and Paredes & Encinas, 2020).

During this period, some actions to fight illegality in the Amazon were taken, the most notable being *Operación Mercurio*, an unprecedented operation launched in 2019 involving a broad coalition of national and local government agencies to (a) aggressively pursue suspected illegal miners in Madre de Dios, (b) rescue victims of human trafficking by criminals associated with mining, (c) ensure local security with the establishment of military bases in the region, and then (d) provide sustainable and diversified employment alternatives as well as creating conditions for local artisanal miners to perform sustainable mining (DuPée, 2019). Between February 2019 and May 2020, gold mining deforestation declined by more than 90% in La Pampa (Madre de Dios), which was the highest intensity illegal mining zone in Peru (MAAP, 2020b). Another notable event is the arrest of regional officials in the Amazon by the anti-corruption police and the criminal prosecutor for alleged criminal activity linked with the illegal granting of land ownership titles (Sierra Praeli, 2018).

Simultaneously, however, MINEM continued to hand out mining concessions in forest areas (and at the beginning of 2020, a figure with connections to illegal mining was appointed to head the ministry); MINAGRI remained passive to land trafficking and the expansion of agriculture in forest land; and the Ministry of Transport and Communications continued to promote the construction of roads in the Amazon[3] (Dourojeanni, 2019, 2020). Additionally, the critical issue of land use planning remained unaddressed, and rural populations continued to lack financial and technical assistance.

Although the team that led MINAM during this period was formed by some of the people who had worked with Pulgar Vidal in Humala's administration, the ministry continued to display a low profile in a complex and unstable political context, focusing on solid-waste management, recycling and the plastic issue (Interviews 7, 8, 12 and 19). The ministry was also involved in controversy due to a decree transferring OSINFOR from the PCM to MINAM, which was criticized as being an attempt to take the institution's independence away, in light of the fact that Fabiola Muñoz, Minister of Environment during most of this period, had been questioned in 2016, as Executive Director of SERFOR, for having supported the recovery of timber identified as illegal by OSINFOR in favor of agro-export businessmen (Interviews 1 and 19; Sierra Praeli, 2018). The decision was eventually reversed and OSINFOR returned to the PCM.

At the time of the conclusion of this chapter, in November 2020, the new Congress voted to impeach Vizcarra. It should also be noted that Peru was one of the most affected countries by the COVID-19 pandemic in Latin America. The difficult political and economic context in the country will likely continue to negatively affect the Amazon over the next few years.

Notes

1 Information provided by a member of the *Instituto del Bien Común*, a Peruvian non-profit civil association working with rural communities.
2 Ibid.
3 Over the past eight years, the extension of the road network in the Peruvian Amazon has more than doubled (RAISG, 2020).

References

AIDESEP (2019, October 29). Pronunciamento sobre malversación de fondos en AIDESEP. Retrieved from http://www.aidesep.org.pe/noticias/pronunciamiento-sobre-malversacion-de-fondos-en-aidesep

Andina (2017, October 3). MEF inició mesa ejecutiva forestal para elevar competitividad. Retrieved from https://andina.pe/agencia/noticia-mef-inicio-mesa-ejecutiva-forestal-para-elevar-competitividad-685071.aspx

Arce, M., & Incio, J. (2018). Perú 2017: un caso extremo de gobierno dividido. *Revista de Ciencia Política*, *38*(2), 361–377. doi:10.4067/s0718-090x2018000200361

Barrenechea, R. (2014). Perú 2013: la paradoja de la estabilidad. *Revista de Ciencia Política*, *34*(1), 267–292. doi:10.4067/S0718-090X2014000100013

Bennett, A., Ravikumar, A., & Paltán, H. (2018). The Political Ecology of Oil Palm Company-Community Partnerships in the Peruvian Amazon: Deforestation Consequences of the Privatization of Rural Development. *World Development, 109*, 29–41. doi:10.1016/j.worlddev.2018.04.001

Bernet Kempers, E. (2020). Between Informality and Organized Crime: Criminalization of Small-Scale Mining in the Peruvian Rainforest. In Y. Zabyelina & D. van Uhm (Eds.), *Illegal Mining: Organized Crime, Corruption, and Ecocide in a Resource-Scarce World* (pp. 273–298). Cham: Palgrave Macmillan.

Caballero Espejo, J., Messinger, M., Román-Dañobeytia, F., Ascorra, C., Fernandez, L.E., & Silman, M. (2018). Deforestation and Forest Degradation Due to Gold Mining in the Peruvian Amazon: A 34-Year Perspective. *Remote Sensing, 10*, 1903. doi:10.3390/rs10121903

Castillo Castañeda, P. (2016). El gobierno de Ollanta Humala y la continuación del síndrome de «el perro del hortelano». *La Revista Agraria, 181*, 10–13.

Charpentier, J. (2016, December 5). Peru Pledges Tougher Stance Against Illegal Timber. *Mongabay Latam*. Retrieved from https://news.mongabay.com/2016/12/peru-pledges-tougher-stance-against-illegal-timber/

Chirif, A. (2012). Aidesep, una crisis de principios. *Ideele, 215*. Retrieved from https://revistaideele.com/ideele/content/aidesep-una-crisis-de-principios

Cooperacción (2014). Sexto informe cartográfico sobre concesiones mineras en el Perú. Lima.

Crabtree, J., & Durand, F. (2017). *Peru: Elite Power and Political Capture*. London: Zed Books.

Dammert, J. L. (2018). Land Trafficking: Agrobusiness, Titling Campaigns and Deforestation in the Peruvian Amazon. Paper presented to the 2018 World Bank Conference on Land and Poverty, Washington DC, 19–23 March. Retrieved from https://www.oicrf.org/-/land-trafficking-agribusiness-titling-campaigns-and-deforestation-in-the-peruvian-amazon

De Echave, J. (2016). La minería ilegal en Perú: entre la informalidad y el delito. *Nueva Sociedad, 263*, 131–144.

De Sy, V., Herold, M., Achard, A., Beuchle, R., Clevers, J.G.P., Lindquist, E., & Verchot, L. (2015). Land Use Patterns and Related Carbon Losses Following Deforestation in South America. *Environmental Research Letters, 10*, 124004. doi:10.1088/1748-9326/10/12/124004

Dourojeanni, M. (2019, November 4). Amazonía: discurso y práctica gubernamental. *SPDA Actualidad Ambiental*. Retrieved from https://www.actualidadambiental.pe/opinion-amazonia-discurso-y-practica-gubernamental-por-marc-dourojeanni/

Dourojeanni, M. (2015, October 16). Carreteras ilegales y destrucción de la Amazonia: ¿qué passa en Madre de Dios? *SPDA Actualidad Ambiental*. Retrieved from https://www.actualidadambiental.pe/construccion-de-vias-ilegales-y-destruccion-de-la-amazonia-que-pasa-en-madre-de-dios/

Dourojeanni, M. (2020, March 4). Irracionalidad de la minería en la

Amazonía peruana. *SPDA Actualidad Ambiental*. Retrieved from https://www.actualidadambiental.pe/opinion-irracionalidad-de-la-mineria-en-la-amazonia-peruana/

Dourojeanni, M. (2017, June 19). Los presidentes peruanos (1960–2016) y la conservación de la naturaleza. *SPDA Actualidad Ambiental*. Retrieved from https://www.actualidadambiental.pe/opinion-los-presidentes-peruanos-1960-2016-y-la-conservacion-de-la-naturaleza/

Dourojeanni, M., Ráez Luna, E., & Valle-Riestra, E. (2016). *Ambiente y recursos naturales en el Perú: quinquenio 2011–2016*. Lima: Derecho, Ambiente y Recursos Naturales.

DuPée, M. C. (2019, March 21). Peru's Militarized Response to Illegal Mining Isn't Enough to Protect the Amazon. *World Politics Review*. Retrieved from https://www.worldpoliticsreview.com/articles/27679/peru-s-militarized-response-to-illegal-mining-isn-t-enough-to-protect-the-amazon

Durand, F. (2016). *Extractives Industries and Political Capture: Effects on Institutions, Equality, and the Environment*. Lima: Oxfam.

Durand, F. (2010). *La mano invisible en el Estado*. Lima: Fondo Editorial del Pedagógico San Marcos.

Eguren, F. (2016). ¿Cómo le fue al agro en estos cinco años? *La Revista Agraria*, *181*, 4–9.

El Peruano (2019, October 2019). Aprueban documento técnico denominado "trazabilidad de los recursos forestales maderables". Resolución de dirección ejecutiva n.° 230–2019-MINAGRI-SERFOR-DE. Retrieved from https://sinia.minam.gob.pe/normas/aprueban-documento-tecnico-denominado-trazabilidad-recursos-forestales

Erickson-Davis, M. (2018, January 18). 680000 Acres of Amazon Rainforest May Be Lost to Peru's New Roads. *Mongabay Latam*. Retrieved from https://news.mongabay.com/2018/01/680000-acres-of-amazon-rainforest-may-be-lost-to-perus-new-roads/

Finer, M., Jenkins, C.N., Blue Sky, M.A., & Pine, J. (2014). Logging Concessions Enable Illegal Logging Crisis in the Peruvian Amazon. *Scientific Reports*, *4*, 4719. doi:10.1038/srep04719

Finer, M., & Orta-Martínez, M. (2010). A Second Hydrocarbon Boom Threatens the Peruvian Amazon: Trends, Projections, and Policy Implications. *Environmental Research Letters*, *5*, 014012. doi:10.1088/1748-9326/5/1/014012

Fraser, B. (2011, October 11). High Gold Price Triggers Rainforest Devastation in Peru. *Mongabay Latam*. Retrieved from https://news.mongabay.com/2011/10/high-gold-price-triggers-rainforest-devastation-in-peru/

Glinskis, E.A., & Gutiérrez-Vélez, V.H. (2019). Quantifying and Understanding Land Cover Changes by Large and Small Oil Palm Expansion Regimes in the Peruvian Amazon. *Land Use Policy*, *80*, 95–106. doi:10.1016/j.landusepol.2018.09.032

Guidi, R. (2016). Peru Terminates Head of Forest Watchdog Agency. *Mongabay Latam*. Retrieved from https://news.mongabay.com/2016/01/peru-terminates-head-of-forest-watchdog-agency/

Hill, D. (2016, May 1). Gold-mining in Peru: Forests Razed, Millions Lost, Virgins Auctioned. *The Guardian*. Retrieved from https://www.theguardian.com/environment/andes-to-the-amazon/2016/may/01/gold-mining-in-peru-forests-razed-millions-lost-virgins-auctioned

Interagency Committee on Trade in Timber Products from Peru (2016, August 17). Statement Regarding July 2016 Timber Verification Report from Peru. Retrieved from https://ustr.gov/sites/default/files/Timber-Committee-Report-8172016.pdf

Ipenza Peralta, C.A. (2018). *Manual de delitos ambientales: una herramienta para operadores de justicia ambiental*. Lima: Derecho, Ambiente y Recursos Naturales.

La Revista Agraria (2009). Cuarenta años después de la reforma agraria, la concentración de la propiedad de la tierra reaparece. *107*, 8–19.

La Revista Agraria (2011a). 2010: mucho ruido y pocas nueces para el sector agrario. *125*, 11–12.

La Revista Agraria (2011b). Un gobierno de Gana Perú, ¿en qué puede ser diferente? *130*, 6–10.

Legislación Ambiental (n. d.). Descentralización e institucionalidad forestal. Retrieved from http://www.legislacionambientalspda.org.pe/index.php?option=com_content&view=article&id=174&Itemid=4065

Leyva, A. (2012). Ordenar el territorio: una voluntad política que se desvanece. *Ideele, 215*. Retrieved from https://revistaideele.com/ideele/content/ordenar-el-territorio-una-voluntad-pol%C3%ADtica-que-se-desvanece

López Tarabochia, M. (2017, August 2). Five Instances in Which Peru Won the Battle Against Deforestation. *Mongabay Latam*. Retrieved from https://news.mongabay.com/2017/08/five-instances-in-which-peru-won-the-battle-against-deforestation/

Lust, J. (2019). *Capitalism, Class and Revolution in Peru, 1980–2016*. Cham: Palgrave Macmillan.

MAAP (2018, April 2). MAAP #81: Carbon Loss from Deforestation in the Peruvian Amazon. Retrieved from https://maaproject.org/2018/peru-carbon/

MAAP (2019, January 8). MAAP #96: Gold Mining Deforestation at Record High Levels in Southern Peruvian Amazon. Retrieved from https://maaproject.org/2019/peru-gold-mining-2018/

MAAP (2020a, October 26). MAAP #127: Mennonite Colonies Continue Major Deforestation in Peruvian Amazon. Retrieved from https://maaproject.org/2020/mennonites_peru/

MAAP (2020b, June 26). MAAP #121: Reduction of Illegal Gold Mining in the Peruvian Amazon. Retrieved from https://maaproject.org/2020/gold-mining/

MAAP (2020c, November 25). MAAP #128: United Cacao Case – 7 Years After Massive Deforestation in the Peruvian Amazon. Retrieved from https://maaproject.org/2020/cacao-tamshiyacu/

Marquardt, K., Pain, A., Bartholdson, O., & Rengifo, L.R. (2019). Forest Dynamics in the Peruvian Amazon: Understanding Processes of Change. *Small-scale Forestry*, *18*, 81–104. doi:10.1007/s11842-018-9408-3

McNulty, S. (2017). Peru 2016: Continuity and Change in an Electoral Year. *Revista de Ciencia Política*, *37*(2), 563–587. doi:10.4067/s0718-090x2017000200563

MEF (2019, July 19). Ejecutivo aprobó medidas para impulsar el desarrollo del sector forestal. Retrieved from https://www.mef.gob.pe/es/notas-de-prensa-y-comunicados/6076-ejecutivo-aprobo-medidas-para-impulsar-el-desarrollo-del-sector-forestal

Meléndez, C. (2009). Perú 2008: el juego de ajedrez de la gobernabilidad en partidas simultáneas. *Revista de Ciencia Política*, *29*(2), 591–609. doi:10.4067/S0718-090X2009000200016

Merino, R. (2019). Rethinking Indigenous Politics: The Unnoticed Struggle for Self-Determination in Peru. *Bulletin of Latin American Research*, *39*(4), 413–528. doi:10.1111/blar.13022

MINAM (2011). Decálogo Ambiental. Lima.

MINAM (2016). La conservación de bosques en el Perú (2011–2016): conservando los bosques en un contexto de cambio climático como aporte al crecimiento verde. Retrieved from http://www.minam.gob.pe/informessectoriales/wp-content/uploads/sites/112/2016/02/11-La-conservaci%C3%B3n-de-bosques-en-el-Per%C3%BA.pdf

Mujica, J. (2014). El *lobby* en un escenario de agendas fragmentadas: consideraciones sobre los mecanismos de gestión de intereses en el parlamento peruano. *Revista de Ciencia Política y Gobierno*, *1*(1), 37–54.

Muñoz, P., & Guibert, Y. (2016). Perú: el fin del optimismo. *Revista de Ciencia Política*, *36*(1), 313–338. doi:10.4067/S0718-090X2016000100014

Observatorio de Conflictos Mineros en el Perú (2010). Sétimo Informe. Retrieved from http://www.spaciolibre.pe/wp-content/uploads/2010/12/OCM_Setimo_informe.pdf

Pachico, E. (2012, April 17). Drug Traffickers Take Norte of Peru's Illegal Timber Trade. *Insight Crime*. Retrieved from https://www.insightcrime.org/news/analysis/drug-traffickers-take-note-of-perus-illegal-timber-trade/

Paredes, M., & Encinas, D. (2020). Perú 2019: crisis política y salida institucional. *Revista de Ciencia Política*, *40*(2), 483–510. doi:10.4067/S0718-090X2020005000116

Peinhardt, C., Kim, A. A., & Pavon-Harr, V. (2019). Deforestation and the United States-Peru Trade Promotion Agreement. *Global Environmental Politics*, *19*(1), 53–76. doi:10.1162/glep_a_00498

Piu Deza, H. C., & Galván Gildemeister, O. (2015). *La transformación del bosque: titulación de predios y cambio de uso de suelos en la Amazonía peruana*. Lima: Derecho, Ambiente y Recursos Naturales.

Ponce de León, Z., & García Ayala, L. (2019). Perú 2018: la precariedad política en tiempos de Lava Jato. *Revista de Ciencia Política*, *39*(2), 341–365. doi:10.4067/S0718-090X2019000200341

Programa de Política y Gobernanza Ambiental de la SPDA (2017, October 6). Minam reduce estándares ambientales para proyectos de exploración minera. *SPDA Actualidad Ambiental.* Retrieved from https://www.actualidadambiental.pe/opinion-minam-reduce-estandares-ambientales-para-proyectos-de-exploracion-minera/

Ráez Luna, E.F. (2019). *La Amazonía peruana y el cambio climático.* Lima: Movimiento Ciudadano frente al Cambio Climático.

Ráez Luna, E. (2020). *Visiones, prioridades y urgencias del Perú: ante la emergencia climática global.* Lima: Movimiento Ciudadano frente al Cambio Climático.

RAISG (2020). *Amazonía bajo presión.* São Paulo: ISA – Instituto Socioambiental.

Ravikumar, A., Sears, R.R., Cronkleton, P., Menton, M., & Pérez-Ojeda del Arco, M. (2016). Is Small-Scale Agriculture Really the Main Driver of Deforestation in the Peruvian Amazon? Moving Beyond the Prevailing Narrative. *Conservation Letters, 10*(2), 170–177. doi:10.1111/conl.12264

Romero, E. (2014). Industria maderera y redes de poder regional en Loreto. *Revista Argumentos, 8*(3), 27–33.

Salazar, B. (2016). La gestión ambiental durante el gobierno de Ollanta Humala. *La Revista Agraria, 181*, 14–17.

Sánchez-Cuervo, A. M., Santos de Lima, L., Dallmeier, F., Garate, P., Bravo, A., & Vanthomme, H. (2020). Twenty Years of Land Cover Change in the Southeastern Peruvian Amazon: Implications for Biodiversity Conservation. *Regional Environmental Change, 20*, 8. doi:10.1007/s10113-020-01603-y

Servindi (2017, May 22). La historia del debilitamiento del ministerio del ambiente (2008–2017). Retrieved from https://www.servindi.org/actualidad-noticias/20/05/2017/la-historia-del-debilitamiento-del-ministerio-del-ambiente-2008-2017

Servindi (2010, May 27). Perú: piden a Alan García promulgar ley de consulta aprobada por congreso. Retrieved from https://www.servindi.org/actualidad/26296

Shanee, N., & Shanee, S. (2016). Land Trafficking, Migration, and Conservation in the "No-Man's Land" of Northern Peru. *Tropical Conservation Science, 9*(4), 1–16. doi:10.1177/1940082916682957

Sierra Praeli, Y. (2018, December 18). Balance ambiental de Perú en 2018: tráfico de tierras, áreas protegidas bajo amenaza y políticas cuestionadas. *Mongabay Latam.* Retrieved from https://es.mongabay.com/2018/12/balance-ambiental-peru-2018/

SPDA Actualidad Ambiental (2011, April 1). Madre de Dios: federación de mineros exige la derogación de decretos que permiten la destrucción de dragas. Retrieved from https://www.actualidadambiental.pe/madre-de-dios-federacion-de-mineros-exige-la-derogacion-de-decretos-que-permiten-la-destruccion-de-dragas/

Valle-Riestra, E. (2017). Los efectos del 'paquetazo ambiental': avances y retocesos en el proceso de ordenamiento territorial y fiscalización ambiental en Perú. *Revista Latinoamericana de Derecho y Políticas Ambientales, 5*(5), 37–47.

Van der Valk, N., Bisschop, L., & van Swaaningen, R. (2020). When Gold Speaks, Every Tongue Is Silent: The Thin Line Between Legal, Illegal, and Informal in Peru's Gold Supply Chain. In Y. Zabyelina, & D. van Uhm (Eds.), *Illegal Mining: Organized Crime, Corruption, and Ecocide in a Resource-Scarce World* (pp. 299–327). Cham: Palgrave Macmillan.

Vergara, A., & Watanabe, A. (2019). Presidents without Roots: Understanding the Peruvian Paradox. *Latin American Perspectives*, *46*(5), 25–43. doi:10.1177/0094582X19854097

Vijay, V., Reid, C.D., Finer, M., Jenkins, C.N., & Pimm, S.L. (2018). Deforestation Risks Posed by Oil Palm Expansion in the Peruvian Amazon. *Environmental Research Letters*, *13*, 114010. doi:10.1088/1748-9326/aae540

List of Interviews

1. Interview with former senior official of MINAM, WhatsApp, July 2020.
2. Interview with former official of MINAM, Microsoft Teams, July 2020.
3. Interview with Peruvian member of an international ENGO, Google Meet, July 2020.
4. Interview with current official of MINAM, Google Meet, July 2020.
5. Interview with current senior official of MINAM, Google Meet, July 2020.
6. Interview with former official of MINAGRI, Google Meet, July 2020.
7. Interview with Peruvian environmental lawyer, Google Meet, July 2020.
8. Interview with Peruvian forest engineer, Google Meet, July 2020.
9. Interview with Peruvian geographer, Google Meet, July 2020.
10. Interview with Peruvian forest engineer, Google Meet, July 2020.
11. Interview with Peruvian member of an international ENGO, Google Meet, July 2020.
12. Interview with Peruvian environmental lawyer, Google Meet, July 2020.
13. Interview with Peruvian environmental lawyer, WhatsApp, July 2020.
14. Interview with member of a Peruvian ENGO, Google Meet, July 2020.
15. Interview with Peruvian environmental lawyer, Google Meet, July 2020.
16. Interview with Peruvian climate scientist, Google Meet, July 2020.

17. Interview with Peruvian member of an international ENGO, Google Meet, July 2020.
18. Interview with Peruvian negotiator of the US–Peru Trade Promotion Agreement, Google Meet, August 2020.
19. Interview with Peruvian member of an international ENGO, Google Meet, August 2020.
20. Interview with former senior official of MINAM, Google Meet, August 2020.
21. Interview with former official of SERFOR, Google Meet, August 2020.
22. Interview with indigenous member of a national ENGO, WhatsApp, September 2020.

4 Bolivia: Between the Rights of Mother Earth and Agro-Extractivism

Bolivia is the world's twelfth most biodiverse country in the world and accounts for over 8% of the Amazon. Deforestation data for the Bolivian Amazon point in rough terms to relatively stable (high) deforestation levels until the late 2000s, followed by a decline in the beginning of the 2010s, and an increase since 2013 (Appendix A, Figure A.3). These are the deforestation trends that we assume in our analysis in this chapter. The perception of most interviewees of this study is, nevertheless, that deforestation has never ceased to increase during this century. Deforestation data (validated in the field) produced by technicians from the *Observatorio del Bosque Seco Chiquitano* (Maillard, Anívarro, & Flores-Valencia, 2020) for Santa Cruz, which is the most deforested department in the Bolivian Amazon, indicate a sharp increase in the average annual deforestation rates of the last decade compared to the previous ones; however, with the exception of 2018–2019, the data were collected and aggregated into large time intervals (1986–2010 and 2010–2018), which precludes a more refined analysis of deforestation tendencies.

Deforestation in the Bolivian Amazon has been mainly associated with (a) the expansion of capital-intensive mechanized agriculture, mostly of soybeans for international markets, often in rotation with sugarcane, rice, sunflower and wheat, by national and international (Brazilian, for the most part) companies, Mennonites and increasingly by settlers; (b) the growth of small-scale agriculture of both subsistence and cash crops mainly for local markets using manual cultivation methods, sometimes combined with livestock activities, to a large extent by settlers coming from the country's highlands; and (c) the expansion of cattle ranching mainly for domestic markets by large-scale cattle ranchers, in some cases with Brazilian capital near the border (Müller, Pistorius, Rohde, Gerold, & Pacheco, 2013). Since the beginning of the century, there has been a strong advance in cattle ranching vis-à-vis other main drivers of land use change – estimates indicate that approximately 60% of deforestation was caused by cattle ranching in the period 2005–2015 (Ferreira, 2019; Müller, Pacheco, & Montero, 2014).[1] Over the past 12 years, Santa Cruz has become the largest hotspot of deforestation in the whole Amazon basin (Kalamandeen et al., 2018).

In the following pages, we analyze the political, economic and social factors that might explain deforestation trends in the Bolivian Amazon between 2006 and 2019, thus covering the entire presidency and three administrations (2006–2009, 2010–2014 and 2015–2019) of Evo Morales and his Movement Toward Socialism (MAS) party. The discussion is divided into two periods: 2006–2012 and 2013–2019. The first period is an ambiguous one, characterized by both progressive pro-environmental regulations and predatory development policies reflecting the existence of opposing factions within the MAS. It is also a period marked by the formation of a constituent assembly to rewrite the Bolivian Constitution toward a plurinational reconfiguration of the state and the launching of an agrarian reform program that earned Morales the fierce opposition of Santa Cruz's elites, putting Bolivia on the edge of civil war. Of particular relevance for the purpose of our study is the fact that the agrarian reform contributed to the high levels of deforestation registered during the first years of the MAS's government. By the end of this period, in a plot twist, the government and Santa Cruz's elites allied. The second period witnessed the consolidation of forces pushing for environmentally-destructive development and the issuing of a series of laws and decrees incentivizing the expansion of the agricultural frontier and violating the country's protected areas, in a context marked by the global fall in commodity prices.

4.1 2006–2012: Land Reform, Elite Contestation and the Clash of Ecology and Development

In December 2005, following multiple popular struggles against neo-liberal policies – with demands for inclusive decision-making processes, sovereignty over national natural resources and a more equitable distribution of Bolivia's resource wealth – as well as deep political instability, Aymara indigenous and cocalero activist Evo Morales was elected president with 54% of the popular vote, supported by a coalition between the country's main indigenous and peasant organizations called the "Unity Pact", and discursively attacking oligarchic and economic elites. The new president, who promised a historic "process of change" that would embrace indigenous emancipatory socioenvironmental demands and break with the West's imperialism, took office in January 2006. In March the same year, a law convoking a constituent assembly to rewrite the national Constitution was passed – the new constitutional text should grant Bolivia's indigenous and peasant peoples direct political representation and powers as collective subjects in conformity with their customary practices (McNelly, 2020). Additionally, in November, a law launching an agrarian reform was enacted. It aimed at accelerating collective land titling, providing land to indigenous peoples and landless communities or peasants, expropriating medium- and large-scale farms and ranches, and ensuring that all land would serve a "socioeconomic function" (FES) – meaning that landowners must make the "best use" of the land to achieve economic development and promote all residents' "best interests" and welfare; otherwise, the land would be expropriated. The new government also nationalized hydrocarbons in 2006 and applied a new direct tax on hydrocarbon resources translating to a 70% cut in the size of revenue-sharing with departments in 2007. Moreover, to promote food security and sovereignty, goods such as sugar and rice were the subject of price controls and the direct sale of soybean oil on the country's urban market was made mandatory in 2008, and restrictions on the export of agricultural goods introduced in 2009 (Eaton, 2017).

The MAS's attempts to reconfigure the state in favor of indigenous peoples and peasant communities, nationalize the economy and transform the Bolivian rural landscape and class structure were obviously not well-received by the large landowners and economic elites of Santa Cruz, Bolivia's most powerful economic region, where hydrocarbons and agribusinesses are located.[2] A four-year regionalist and autonomist violent destabilization campaign against the government was thus initiated by *cruceño* elites – headed by Rubén Costas, the region's governor and opposition leader – and joined by those in Beni, Pando and Tarija, the other lowland departments of the country. This coalition threatened to tear Bolivia apart; in 2008, there was almost a coup d'état. Nevertheless, a series of false steps – the most serious being the massacre of indigenous peoples by local elites in Pando – and the solid support of social movements of Morales throughout the country, led to the defeat of the opposition's regional autonomy movement (Farthing, 2019; Webber, 2016). Still, to de-escalate violence and because his party did not control the Senate (right-wing parties did), Morales was forced to make a number of important concessions to lowland elites. More than 100 changes in the draft of the new Constitution were made before the text was submitted to, and approved through, a national referendum in 2009, the most significant being a concession on land reform. Although the Constitution set a maximum land-size limit of 50 km^2, this limit was defined as not being retroactively binding for pre-existing large farms and ranches; in addition, the Constitution allowed for agro-industrial groups to comprise an unrestricted number of business associates and authorized ownership of the maximum 50 km^2 limit of land by each of them (McKay, 2018). Morales was also eventually forced to lift price controls and export restrictions on agricultural goods following food shortages across the country as a consequence of the retainment of sugar, rice, oil and meat from the domestic market by Santa Cruz's producers in 2009 – the region plays a preponderant role as the country's provider of food (Eaton, 2017).

It is thus unsurprising that very few of the country's large landowners were expropriated. Most of the land titled in the context of the agrarian reform was endowed by the state to lowland indigenous peoples in the form of native community lands, or to peasants and syndicates, and consisted of state-owned forest land. Many encroached on public lands with the expectation of ownership being granted under the reform, and to comply with the FES requirement, converted forests into agriculture and pasture land. Others did the same, fearing expropriation (Müller et al., 2014; Pacheco & Benatti, 2015; Redo, Millington, & Hindery, 2011). As a result, deforestation levels remained high at least during the first years of Morales's administration. It should be noted, nonetheless, that over the next few years, the government kept distributing available state-owned land suitable for forestry to Morales's supporters, particularly highland Aymara and Quechua migrants (Webber, 2016).

Another factor that contributed to forest clearing during this period was the displacement of deforestation processes from Brazil to Bolivia, as a result of the ambitious, strong deforestation control policy implemented in the Brazilian Amazon between 2005 and 2010 (Interviews 1, 7 and 8; Kalamandeen et al., 2018), as seen in Chapter 2.

What could explain the deforestation decline in the following years, as suggested by the various data sources consulted? In 2008, the national policy for comprehensive forest management was designed to reorient public forest policy in the country. It provided the basis for developing policy proposals for the forest sector, reforming the institutional system and gradually adjusting regulations to improve forest control and the state's sanctioning capacity. Within this context, the Forest and Land Authority (ABT) – a governmental agency to oversee land use and change – was created in 2009 (Müller et al., 2014). It was initially attached to the Ministry of Rural Development and Land, but was transferred to the Ministry of Environment and Water (MMAyA) in February 2010 (ABT, 2012). ABT's Director, Clíver Rocha, implemented a number of administrative processes to control deforestation and illegal logging. The institution's controlling, supervising and sanctioning activities, with a significant increase in the value of fines for illegal forest clearing, could explain a potential reduction in deforestation levels at the beginning of the 2010s[3] (Interviews 2 and 3; ABT, 2018; Cámara Forestal de Bolivia, 2014; Unitel, 2014).

Other relevant legislative and programmatic initiatives were launched during this period, such as the national strategy for forestry and climate change; a law (071/2010) recognizing the rights of Mother Earth and another (300/2012) establishing the principles for harmonizing the integral development of the Bolivian people with environmental protection, under the indigenous ethical and philosophical principles of "*Vivir Bien*" ("Living Well");[4] and an alternative mechanism to REDD – which was criticized by the most radical ecologist factions of the government for its market approach – called the Joint Mitigation and Adaptation Mechanism for the Integrated and Sustainable Management of Forests and Mother Earth, based on the non-marketization of the environmental functions of nature (Müller et al., 2014). It should also be noted that it was the Morales administration that granted political autonomy to the environmental sector by creating the MMAyA in 2009; previously, the sector was integrated into the Ministry of Rural Development, Agriculture and Environment.

Simultaneously, however, particularly from its second mandate, the Morales government was deepening the country's predatory, resource-extraction development model. The resistance of the country's economic elites was not the only challenge faced by the MAS; the party also struggled with the contradictory stances on issues regarding lands, territories and resources that existed within itself. Many of its members and supporters had vague or even conflicting ideas about what the party's core ideology was or should be; they were unified around their common rejection of neoliberal policies and the status quo as well as their condition as historically marginalized citizens (McKay, 2018). The Unity Pact, for example, comprised two types of actors with very different views and interests – peasant unions, whose conception of land is one of individual ownership and intensive agricultural use associated with market-oriented production, and indigenous organizations emphasizing subsistence economy practices based on communitarian property of land and territorial self-governance. Peasants have argued that, while they have remained forgotten, indigenous communities, particularly those of the Amazon, had been overly assisted by NGOs and foreign aid agencies since the 1990s, and favored by neoliberal governments, having had great advances in their collective land rights; they have thus called for an even and equal distribution of land. The MAS's policy of granting land titles to its highland supporters has created tensions between Amazonian communities and settlers. There was also a broader division between, on the one hand, the party's indigenous and ecologist bloc, and on the other hand, the "traditional Left" faction, strongly represented by Vice-President Álvaro García Linera, which prevailed from Morales's second mandate. This faction advocated for a state-led developmentalism and resource nationalism (in other words, the defense of a deepening of the actions of transnational companies but by state enterprises, with revenue redistribution), and considered indigenous demands an obstacle for the nation's development. During the MAS's years in power, there were significant advances in reducing poverty and inequality, and therefore widespread support for extractivism as a means of improving social conditions. The rights of Mother Earth and those of Amazonian indigenous peoples collided with broader social rights and those of the state to extract and commercialize the country's natural resources under the banners of social reform, redistributive justice and the common good (see, for instance, Laing, 2020; Lalander, 2017; Ranta, 2018; Springerová & Vališková, 2016). For García Linera (2012), extractivism was a means of overcoming extractivism itself in the long-term.

Tensions escalated in 2011, due to the government's plan of building a highway through the Isiboro Securé National Park and Indigenous Territory (TIPNIS). The highway would facilitate the transportation of agricultural goods from the Amazon to wider markets and create new economic opportunities for Morales's main political support bases, that is, Aymara and Quechua settlers, and coca-growing peasants (in many cases, for cocaine production and trafficking). Indigenous groups, fearing that the new highway would further enhance migration to the area and accelerate deforestation, and joined by environmental and human rights NGOs, initiated a protest march, and faced opposition by peasant unions – which remained loyal to the government and organized a road blockade – and brutal police repression. For García Linera, those who opposed the highway were at the service of the country's right-wing opposition and foreign imperialism; they were "colonial environmentalists" financed by industrialized countries, who were denying the right to development of the Bolivian people (Ranta, 2018; see also Colque, 2018; García Linera, 2011; 2012). The TIPNIS conflict led to the breaking of the alliance between indigenous movements and peasant unions. To reduce the indigenous threat to resource-based accumulation, the government weakened the anti-government indigenous organizations. As explained by Andreucci (2018, p. 837):

> [I]t divided and disarticulated the main indigenous organisations, in order to control them. Commencing in 2012 with the lowland indigenous federation, CIBOD (Confederation of Indigenous Peasants of Bolivia), the government identified members and cadres aligned with the party, or willing to be co-opted. It created parallel organizations under the control of the MAS and isolated and marginalised the remaining – that is, legitimately elected – leadership members. This formed a slip between pro-government (MASista) and independent (*organico*) organisations. A similar pattern was followed (...) with (...) CONAMAQ [National Council of *Allyus* and *Markas* of the *Qullasuyu*]. This allowed for the formal reconstitution of the Unity Pact, this time fully controlled by the government.

In the words of McNelly (2020, p. 88), organizations became "empty shells" and "ceased to be movements".

As for protected areas, there were also severe setbacks. At the beginning of its mandate, the Morales administration appointed a new director to the National Service of Protected Areas (SERNAP) and replaced the majority of the institution's technicians with people close to the MAS, which created instability within the institution and severely affected its capacity. Consequently, there was an unprecedented de-structuration and inactivity in the management of the country's protected areas (Miranda Chávez, 2016). For the environmentally-conservative part of the government, protected areas were an invention of the North American empire and were hindering the country's development; the idea was that Bolivia must recover sovereignty over its forests (Gautreau & Perrier Duslé, 2019). Nevertheless, following protests against the de-institutionalization of SERNAP, the government appointed an indigenous leader from TIPNIS to head the institution. A new model of management for protected areas including indigenous and peasant communities, whose territories in many cases overlapped with protected areas, was designed within the dynamic of the still intact Unity Pact and in the context of the constituent assembly – the "territorial management with shared responsibility" (GTRC) model, whose aim would be to promote both environmental conservation and the territorial rights of indigenous peoples. A proposal for a decree on the GTRC was presented and unsurprisingly rejected in 2010 by the executive, whose predatory development policies were at odds with the conservation of protected areas. In 2012, the GTRC's unit within SERNAP ceased its functions and another indigenous from TIPNIS, who was aligned with settlers and coca-growing peasants and therefore with the government, was designated to lead the institution. The technical capacity recovered under the previous administration was lost, and SERNAP became once more dominated by conservative supporters of the MAS. Cooperation with international bodies was broken, and NGOs working in the country were evicted or became strictly controlled, with their role in the forestry sector being greatly reduced (Gautreau & Perrier Buslé, 2019; Miranda Chávez, 2016). The GTRC proposal was permanently frozen, and the absence of institutional guidelines for allowing social participation in protected area management led to the stagnation of the constructive co-management model of protected areas that existed in the country, through which civil society groups participated in protected area management (Miranda Chávez, 2016; see also Mason, Baudoin, Kammerbauer, & Lehm 2010). It should also be noted that the implementation of the REDD program in Bolivia stopped completely; the program's funds would be channelled to the already mentioned alternative Joint Mechanism on forests, but this had little support from some public, private and community sectors, clear positions on the proposal did not arise, and its procedures were not explicit – as a result, implementation was left pending (Müller et al., 2014). The Morales administration stance caused the country to lose critical opportunities for attracting international funding and partnerships for forest protection, at a time when those opportunities were flourishing, as seen in the chapters on Brazil, Peru and Colombia. There were also few advances in the creation of national protected areas during Morales's presidency; against this background, subnational governments took the lead and created their own protected areas at departmental and municipal levels (Interviews 3, 5, 6, 9 and 10).

It is important to note that, after the 2009 general elections, Morales won with 64% of the popular vote and the MAS gained control of both the Chamber of Deputies and the Senate, which provided an opportunity for pursuing structural reforms. However, what the government made instead was an alliance with the economic elites of Santa Cruz. Given their economic and political influence, and the region's major role as a provider of foods, as well as the slowdown in commodity prices since 2011 and the mutual interests of both the elites and the government, dominated by the "traditional Left" faction, in the expansion of resource extraction, the latter opted for strategically incorporating the *cruceños* into its predatory development model. The elites of Santa Cruz also came to the conclusion that seeking a good relationship with the Morales administration could serve their interests better than opposing it, given the MAS's broad popular support and the rapid economic growth happening under its government, which could bring them new economic opportunities; in fact, during the Morales years in power, the region became one of the fastest-growing in the world (Eaton, 2017; Farthing, 2019; McKay, 2017; McNelly, 2020). In 2011, a law (144/2011) aimed at establishing rules for what the government called the "productive agricultural community revolution" for food sovereignty was issued; nevertheless, although it "guide[d] interventions in agricultural matters towards community organizations, in practice it rather implie[d] the promotion of agro-industrial activities, not directly, but through legal provisions that facilitate[d] development by making regulations more flexible, mainly those related to forest clearing" (Müller et al., 2014, p. 31). This law marked the beginning of the rapprochement between the Morales administration and *cruceño* economic elites. This alliance consolidated the rupture between the government and the Amazonian indigenous organizations in the context of the TIPNIS conflict, and developed

> with subordinate support from (...)[some groups of] peasants [particularly highland migrant Aymara and Quechua peoples] in the coca, soy, and quinoa commercial export sectors, among others (...)[, who have been] incorporated economically into wider, transnationally controlled commercial crop value chains, and politically into the ruling bloc agrarian coalition between the state and industrial agricultural capital (Webber, 2016, p. 331).

Those peasants were deeply integrated into the state under the Morales administration, having direct or indirect influence over a series of public institutions, such as the National Service of Agrarian Reform (INRA), the National Agrarian Commission (CAN) and the Ministry of Rural Development and Land (Carvajal, 2019; Webber, 2016). We will return to the topic of the predatory agrarian coalition formed during the Morales administration in the next subsection.

The alliance with Santa Cruz, in addition to the diminishing autonomous political activity of social movements, with not only their co-optation from above (initially under the idea of decolonizing the state by giving political voice to those who had been marginalized), but also demands by the movements themselves for being incorporated into the government; the creation of parallel social organizations dominated by the MAS and political clientelism practices with pro-government groups receiving considerable funds as well as social welfare cash transfers programs (e.g., *Renta Dignidad* and *Bono Juancito Pinto*), consolidated the MAS's hegemony and legitimacy, and cleared the way for the implementation of the government's extractivist policies (see Farthing, 2019; McKay, 2017; McNelly, 2020; Salazar Lohman, 2020).

The legal recognition of the rights of Mother Earth and the adoption of the *Vivir Bien* philosophy as the search for a life in harmony with others, both human and non-human, freed from the pressures that savage Western capitalism and consumerism put on people, had a symbolic and strategic value for the government internationally. On a symbolic level, it represented the need to find alternatives to predatory capitalism and novel responses to the climate crisis, and transformed Bolivia into the global leader in the pursuit of such alternatives – Evo Morales, who was referred to as a hero in the fight against climate change around the world, and David Choquehuanca, the indigenous Minister of Foreign Affairs between 2006 and 2016, were the most visible faces of that alternative discourse on an international level. However, what the government aimed to do was bring the discussion to international fora and propose the *Vivir Bien* philosophy to the world; on a national level, in a context marked by poverty and inequality, Morales had a much different approach. On a strategic level, the radical socioenvironmental discourse gave the government international prestige and legitimacy, and diverted attention away from the authoritarian, repressive and violent actions perpetrated in the country against those who opposed extractivism. Moreover, from the rupture with its most progressive social bases and the alliance it made with economic elites, the government (and even the elites themselves) began to use the idea of *Vivir Bien* and demands for decolonizing the state to legitimize control over territories and resource-based accumulation, thus dismantling the original progressive indigenous and ecologist meaning of "living well" (Interviews 2, 3, 5, 6, 7, 9 and 10; see also, for instance, Lalander, 2017; Ranta, 2018).

In sum, deforestation during this period was mainly the result of (a) the agrarian reform put through by the Morales government to respond to demands of his social support bases, which translated into a rush for land and during which, mainly forest state land was titled (as it was not possible to recover large quantities of land from large landowners for redistribution), with the FES requirement incentivizing forest clearing by those seeking to obtain ownership and those aimed to avoid expropriation; and (b) the displacement of deforestation processes from Brazil to Bolivia. The alliance that the Morales government made with the economic elites of Santa Cruz by the end of this period, with Law 144/2011 being the first manifestation of such coalition, may have not translated into an increase in deforestation at the beginning of the 2010s due to ABT's supervisory action.

4.2 2013–2019: The Consolidation of Predatory Development Forces

During this period, the coalition between the state, agro-industrial groups and privileged peasants consolidated and was particularly visible from 2015, when falling hydrocarbon and mineral prices reduced state revenues substantially, affecting social welfare programs (Alberti, 2016). The government needed new sources of revenue and focused on agro-industry.

In 2013, a law (337/2013) establishing an exceptional regime for unauthorized forest clearing carried out between 1996 and 2011 was enacted and welcomed by the agro-industrial sector. It allowed farmers to regularize illegal deforestation in exchange for the payment of a reduced fine and enrollment in the food production and forest restitution program, which included a commitment to replant 10% of the area that had been cleared and restore ecological reserves (Müller et al., 2014; Villalobos, 2020). Essentially, the law pardoned the agribusiness groups of Santa Cruz for their historic responsibility for deforestation (Gautreau & Perrier Bruslé, 2019). With law no. 337, agro-industry groups and cattle ranchers were able to avoid land reversion and penal sanctions, and save a significant amount of money due to the establishment of relatively low fines for illegal deforestation (Ormachea & Ramirez, 2013). It should also be noted that the obligation contained in the law to reforest part of the area cleared was insufficient to restore landscape connectivity and functionality, or the forest's ecological services (Interview 7); moreover, the application of restoration was very modest, as the law was mainly perceived as a mechanism for legalizing deforestation rather than one for recovering the forest that had been lost (Murcia, Guariguata, Peralvo, & Gálmez 2017). In the following years, the law was modified by laws no. 502/2014, 739/2015 and 952/2017, which extended the deadline for subscribing to the food production and forest restitution program. This meant that the government lengthened the forgiveness granted to forest clearing to December 2017, the deadline established in law no. 952 (Villalobos, 2020).[5]

In 2014, in a meeting with the President of the Agricultural Chamber of the East and Rubén Costas, the government announced plans to expand Bolivia's agricultural surface by 10,000 km^2 over the next decade (McKay, 2020). The same year, the Director of ABT since 2010, Clíver Rocha, who was critical of massive deforestation by cattle ranchers and advocated for a more efficient system of production that reduced the need to clear more forest (Cámara Forestal de Bolivia, 2014), was replaced by Rolf Köhler, for whom the institution should, on the contrary, serve as a tool for promoting the expansion of the agricultural frontier and cattle ranching (Roca, 2018), and whose administration was praised by economic groups for its contribution to norm flexibilization (ABT, 2018). In the following years, corruption grew within the institution (Servindi, 2019).

In March 2015, a supreme decree (2298/2015) reduced prior consultation with indigenous peoples regarding resource extraction and infrastructure building in their territories to a mere administrative procedure; in May, supreme decree no. 2366/2015 allowed hydrocarbon exploration and extraction inside protected areas (Romero-Muñoz et al., 2019). That same year, the agricultural summit "Sowing Bolivia" took place in Santa Cruz and set the targets of expanding food production and increasing exports as well as triple the contribution of agriculture to the country's GDP. The summit was initially planned by the government and the *cruceño* agrobusiness export sector, but was later extended to small and medium producers, given the need to legitimize both the event and its outcomes (Soliz Tito, 2015). The summit consolidated the predatory agrarian coalition and was the starting point for a series of laws and decrees that directly or indirectly incentivized forest clearing and burning, and the expansion of the agricultural frontier (Cortés Martinez, 2019): (a) law no. 740/2015 established an exceptional five-year period in the FES verification process, applicable to procedures for the reversal of agrarian properties; (b) law no. 741/2015 increased from 0.05 km^2 to 0.2 km^2 the forest land area that could be cleared by small producers for agricultural and livestock activities, without any land use planning or comprehensive forest and land management plans – according to most interviewees, this law has been one of the major drivers of deforestation in the country in recent years; (c) law no. 1098/2018 established the normative framework for the production of biofuels and was complemented by supreme decree no. 3874/2019 expanding the agricultural frontier for the production of genetically-modified soybeans; (d) law no. 1171/2019 authorized the use of fire in productive activities and established an exceptional period for regularizing the payment of debts and fines associated with illegal burning – this law is considered a major cause of the 2019 Amazonian fires in Bolivia; and (e) supreme decree no. 3973/2019 authorized land clearing and the "controlled" burning of forests on private and community lands in the departments of Santa Cruz and Beni for agriculture and livestock production, much of it destined to Chinese markets (see Sierra Praeli, 2019a; 2019b; Villalobos, 2020). The intention to change the land use plan of the Amazonian department of Beni, to make it the largest food producer in Bolivia, had already been announced in 2018 by the ABT's Director Rolf Köhler (he himself a *beniano*) (Franco Berton, 2018), who resigned later that year due to health issues. Curiously, Morales asked Köhler's predecessor, Clíver Rocha, to return to the institution and assume once more the role of director, and clean ABT of the corruption that had been installed within the organization over the preceding few years (Los Tiempos, 2019), which he actually tried to do (Servindi, 2019). His appointment was contested by MAS's supporters, who considered Rocha an authoritarian due to his rigid past administration (Notiboliviarural, 2019; Opinión, 2019). Pressure over the ABT was much stronger during Rocha's second administration than it had been in the first (Los Tiempos, 2019); he was removed from office in the context of the Amazonian fire crisis of 2019, during which the ABT committed a series of irregularities that aggravated the situation in the affected areas, such as the issuing of slash-and-burning and land clearing permits after the beginning of the fires, and to which the government in general responded timidly and slowly (CEDIB, 2020).

According to the analysis conducted by Fundación Tierra (2019), the fires were closely linked to the advance in and consolidation of (mostly large-scale) private agricultural properties, cattle properties and community settlements in major extensions of the agricultural and livestock frontier. Moreover, the crisis exposed the existence of hundreds of settling authorizations granted to new peasant communities in Santa Cruz, reflecting the politics of patronage underlying the processes of land regularizing and titling under the MAS regime (Fundación Tierra, 2019; Interviews 3, 7, 8, and 9).

Finally, during Morales's presidency, there was also an unprecedented expansion of roads in areas of high ecological value to promote national integration and provide access to markets to rural producers (Romero-Muñoz et al., 2019).

The MMAyA was too weak to defend the country's environmental interest in the face of strong political and social pressure in favor of predatory development. From the ministry's creation in 2009 to Morales demise from the presidency in 2019, seven ministers headed the institution (Los Tiempos, 2017) amid the deep contradictions between environmental rhetoric and practice (Interviews 3, 5, 6, 7, 8 and 9). Moreover, with the creation of the MMAyA, the environmental sector was merged with the Ministry of Water,[6] thus also incorporating the management of complex, highly conflictive themes related to water governance, including irrigation and sanitation, in a country facing a deepening water crisis. The governance of water resources, particularly the issue of supply infrastructure, was thus the priority of the government and of the ministry itself, having received the bulk of governmental support. The ministry's financial and technical capacity to protect the country's forests and biodiversity (and even manage water from a conservation point of view) was therefore limited (Interviews 4, 6 and 8). Consequently, the MMAyA was unable to build a solid, strategic vision or plan for the environmental conservation sector, and challenge anti-environmentalist norms (Interview 1).

Additionally, as has already been seen, social movements had been severely weakened over the previous few years. In rural Bolivia, key organizations have been demobilized. In the agriculture sector, and the soy complex in particular, there have emerged "new forms of 'productive exclusion' whereby the capital-intensive form of production has excluded smallholders from production without necessarily divorcing them from their land. (…) [This] has impeded overt forms of resistance" (McKay, 2018, p. 1259). Lacking credit and capital, and faced with a highly-mechanized agro-industrial model that requires few labor power, small farmers are, in many cases, no longer agricultural producers. They have been priced out of producing on their lands and have become dependent on renting them to those with the necessary capital to invest in machinery and inputs for production (that is, agribusinesses and rich peasants), receiving 18%–25% of the net profits from cultivation, and/or turned into semi-proletariats (taxi and bus drivers, shopkeepers, construction workers, etc.). Consequently, their interests have aligned with those of agro-industrial groups, as they, too, depend on the development and expansion of the formers' activities[7] (McKay & Colque, 2016; McKay, 2017; Webber, 2016). Smallholders joined the demands of large landowners and agribusinesses, and these justified their demands on the grounds that they were speaking not only in favor of themselves, but also for a mass of smallholders and settlers (Interviews 1 and 9). A rural dynamic of three related sectors has thus emerged:

> First, there is a nucleus that controls the bulk of land rent [large landowners and agro-industrial groups]. Second, there is a semi-peripheral sector the viability of which is ensured through

> subordinated relations with the nucleus [particularly Quechua and Aymara settlers]. Third, and finally, there is a sector that is either functional to the first two in the form of wage labour or a reserve army of potential labour, or a surplus rural population excluded altogether from the requirements of capital accumulation in the countryside (Webber, 2016, p. 344).

Between 2016 and 2019, however, the legitimacy of Morales and his government declined. This was essentially the result of five factors. First, the President ignored the result of a popular vote (narrowly) rejecting a proposal to change the country's constitution and allow him to run for another presidential election in 2019, which triggered a wave of protests, cost Morales the support of traditional, non-indigenous urban middle classes and disappointed parts of the popular sectors. Second, the end of the commodities boom, which affected the development model pursued by the government. Third, a less favorable international context with the rise to power of right-wing governments in several countries in the American continent, the Venezuelan crisis and the cessation of regional organizations politically associated with the Morales administration, such as the Union of South American Nations (UNASUR). Fourth, the 2019 Amazonian fire crisis in the months preceding the October general elections, to which, as already seen, the government responded inadequately (Wolff, 2020), and during which the MAS's contradictory discourse on the rights of Mother Earth, extractivist policies, measures to promote the expansion of the agricultural frontier and alliance with the agro-industrial groups of Santa Cruz were questioned (Brockmann Quiroga, 2020). Fifth, allegations of fraud in the presidential elections – from which Morales emerged victorious – which led to violent confrontations between supporters and opponents of the MAS. Within this context, seeing an opportunity to recover their hegemony, the conservative elites of the lowlands, which were in a mere tactical alliance with the MAS, re-assumed the role of main opposition force against the party. Amid violence, the head of the armed forces "invited" Morales to resign, which he did on November 10. Jeanine Áñez, a conservative opposition senator from the department of Beni, became the country's president on an interim basis, until new elections were held, with a government closely linked to traditional economic elites (Wolff, 2020). Unsurprisingly, the interim administration incentivized the agro-industrial expansion initiated with Morales.

In the elections of October 2020, nevertheless, the MAS returned to power with Luis Arce, who was Minister of Economy during

Morales's regime, as president, and the former Minister of Foreign Affairs, David Choquehuanca, as vice president. The latter, being a representative of the more indigenist, less developmentalist faction of the MAS, could play a critical role in advancing the country's environmental agenda. Nevertheless, the pressure created by the economic crisis triggered by the COVID-19 pandemic in favor of extractivism will most likely be too strong.

Notes

1 This expansion has probably been partly driven by land speculation; however, there is still no clear evidence on the matter (Interview 6; Pacheco Balanza, 2014).

2 For a reading on the rise of Santa Cruz in the Bolivian economic and political landscape, see, for instance, Eaton (2017).

3 However, because of "the forest technical training of ABT's professionals, which is based on marketable species, that is, on the commercial value of the wood, and not on the ecological role that a forest must fulfil or the soil's function, the institution's promotion of agroforestry and silvopastoral management has been marginal since its creation (…). This is a significant weakness in the forest institutional sector in Bolivia" (Interview 5).

4 It should be noted, nevertheless, that in spite of the centrality of nature and indigenous thought in the law, its "integral development" component included a very pragmatic, anthropocentric dimension centered on human values and interests (Lalander, 2017; see also Pacheco Balanza, 2014; Orellana Halkyer, 2019).

5 In their assessment of the Morales administration environmental policy, Romero-Muñoz et al. (2019, Figure 1) classified law no. 337 as a policy representing an advance to nature conservation in Bolivia. The authors do not discuss the law or provide an explanation for considering it a pro-environmental one. It may be because the law also aimed to promote both the use of land already deforested to food production and reforestation. Nevertheless, in our view, and that of most of the interviewees in this study, any law pardoning illegal deforestation sets a dangerous precedent and risks creating a climate of impunity. The 2019 Amazonian fires were in reality the result of the expectations created by the series of governmental anti-environmentalist policies and legislative initiatives of the previous years (Fundación Tierra, 2019). In addition, one cannot forget that the law was supported by the agribusiness groups of Santa Cruz. For all interviewees except one, law no. 337 was a clear setback in forest conservation. The only interviewee (Interview 2) who had a more nuanced opinion of the law mentioned that it aimed at reducing illegal deforestation. However, legalizing the illegal is not a conservation/deforestation control policy. In fact, what the government did was not reduce deforestation, but rather legalize it. Another piece of information provided by the same interviewee was that many refrained from expanding the agricultural frontier until the rules of the law became clear; nonetheless, even if this was the case, that does not make law no. 337 a pro-environmental one.

6 An emblematic and important ministry due to the fierce struggles over access to water that took place in the country. See, for instance, Fabricant and Hicks (2013).

7 It is interesting to note, though, that in the context of the agricultural summit of 2015, the supposedly government-aligned Unity Pact stood by the interests of smallholders, peasants and indigenous peoples, and rejected many of the proposals presented (which favored mainly the interests of agro-industrial groups), as these would have critical, broad implications for issues pertaining to land access, genetically-modified foods, deforestation, etc. for many within the pact (contrary to what happened in the TIPNIS case – an isolated issue in a specific territory). This demonstrated that the pact, more than being a political instrument of the MAS, was a fluid movement of movements, with "internal tensions and debates [that] must be understood in their specific contexts and with the issues at hand" (McKay, 2018, p. 1259).

References

ABT (2018, December 26). Clíver Rocha retoma el timón de la ABT ante el recelo de empresarios. Retrieved from http://abt.gob.bo/index.php?option=com_content&view=article&id=1285:cliver-rocha-retoma-el-timon-de-la-abt-ante-el-recelo-de-empresarios&catid=88&Itemid=101&lang=en

ABT (2012). Información institucional: ABT. Retrieved from http://www.abt.gob.bo/index.php?option=com_content&view=article&id=52&Itemid=260&lang=es

Alberti, C. (2016). Bolivia: La democracia a una década del gobierno del MAS. *Revista de Ciencia Política*, *36*(1), 27–49. doi:10.4067/S0718-090X2016000100002

Andreucci, D. (2018). Populism, Hegemony, and the Politics of Natural Resource Extraction in Evo Morales's Bolivia. *Antipode*, *50*(4), 825–845. doi:10.1111/anti.12373

Brockmann Quiroga, E. (2020). Tentativa de toma gradual del poder: prorroguismo fallido y transiciones. In F. Mayorga (Ed.), *Crisis y cambio político en Bolivia. Octubre y noviembre de 2019 en Bolivia: la democracia en una encrucijada* (pp. 29–60). La Paz: Centro de Estudios Superiores Universitarios de la Universidad Mayor de San Simón (CESU-UMSS).

Cámara Forestal de Bolivia (2014, October 27). Entrevista a Clíver Rocha horas antes de su reemplazo. Retrieved from https://www.cfb.org.bo/noticias/institucional/entrevista-a-cliver-rocha

Carvajal, R. (2019, September 1). Incendios develan la madre oculta del desastre. *Servindi*. Retrieved from https://www.servindi.org/actualidad-informe-especial/01/09/2019/incendios-develan-la-madre-oculta-del-desastre

CEDIB (2020). Los incendios en la Chiquitania el 2019: políticas devastadoras, acciones irresponsables y negligencia gubernamental. Retrieved from https://cedib.org/wp-content/uploads/2020/09/Dossier-Incendios-Chiquitania.pdf

Colque, G. (2018). Los cocaleros en el conflicto del TIPNIS. *Cuestión Agraria, 4*, 125–147.
Cortés Martinez, P. D. (2019). Retrocesos legales en la protección de la Madre Tierra (latifundio y agronegocio). In Fundación Tierra (Ed.), *Conferencia 2018. Madre Tierra: la agenda abandonada. Causas y consecuencias* (pp. 111–118). La Paz: Fundación Tierra.
Eaton, K. (2017). *Territory and Ideology in Latin America: Policy Conflicts Between National and Subnational Governments.* Oxford: Oxford University Press.
Fabricant, N., & Hicks, K. (2013). Bolivia's Next Water War: Historicizing the Struggles over Access to Water Resources in the Twenty-First Century. *Radical History Review, 2013*(116), 130–145. doi:10.1215/01636545-1965757
Farthing, L. (2019). An Opportunity Squandered? Elites, Social Movements, and the Government of Evo Morales. *Latin American Perspectives, 46*(1), 212–229. doi:10.1177/0094582X18798797
Ferreira, W. J. (2019). Estado de situación de la deforestación en Bolivia. In Fundación Tierra (Ed.), *Conferencia 2018. Madre Tierra: la agenda abandonada. Causas y consecuencias* (pp. 97–109). La Paz: Fundación Tierra.
Franco Berton, E. (2018, July 23). San Carlos: una estancia ganadera que conserva la fauna silvestre en Bolivia. *Mongabay Latam.* Retrieved from https://es.mongabay.com/2018/07/san-carlos-estancia-ganadera-fauna-silvestre-bolivia/
Fundación Tierra (2019). Fuego en Santa Cruz: balance de los incendios forestales 2019 y su relación con la tenencia de la tierra. Retrieved from http://www.ftierra.org/index.php/component/attachments/download/194
García Linera, A. (2011). *El "oegenismo", enfermedad infantil del derechismo. (O cómo la "recondución" del Proceso de Cambio es la restauración neoliberal).* La Paz: Vicepresidencia del Estado Plurinacional.
García Linera, A. (2012). *Geopolítica de la Amazonía: poder hacendal-patrimonial y acumulación capitalista.* La Paz: Vicepresidencia del Estado Plurinacional.
Gautreau, P., & Perrier Buslé, L. (2019). Forest Management in Bolivia under Evo Morales: The Challenges of Post-Neoliberalism. *Political Geography, 68*, 110–121. doi:10.1016/j.polgeo.2018.12.003
Kalamandeen, M., Gloor, E., Mitchard, E., Quincey, D., Ziv, G., Spracken, D.…. Galbraith, D. (2018). Pervasive Rise of Small-Scale Deforestation in Amazonia. *Scientific Reports, 8*, 1600. doi:10.1038/s41598-018-19358-2
Laing, A. F. (2020). Re-Producing Territory: Between Resource Nationalism and Indigenous Self-Determination in Bolivia. *Geoforum, 108*, 28–38. doi:10.1016/j.geoforum.2019.11.015
Lalander, R. (2017). Indigeneidad, descolonización y la paradoja del desarrollismo extractivista en el Estado Plurinacional de Bolivia. *Revista Chilena de Derecho y Ciencia Política, 8*(1), 49–83.
Los Tiempos (2017, January 23). Los hombres y mujeres que acompañaron a

Evo en 11 años. Retrieved from https://lostiemposdigital.atavist.com/los-gabinetes-de-evo-morales

Los Tiempos (2019, September 9). Rocha, exdirector de la ABT, revela que hay "fuego cruzado" por quemas. Retrieved from https://www.lostiempos.com/actualidad/pais/20190908/rocha-exdirector-abt-revela-que-hay-fuego-cruzado-quemas

Maillard, O., Anívarro, R., & Flores-Valencia, M. (2020). Pérdida de la cobertura natural (1986–2019) y proyecciones de escenarios a futuro (2050) en el Departamento de Santa Cruz. Informe técnico. Santa Cruz: Fundación para la Conservación del Bosque Chiquitano.

Mason, D., Baudoin, M., Kammerbauer, H., & Lehm, Z. (2010). Co-Management of National Protected Areas: Lessons Learned from Bolivia. *Journal of Sustainable Forestry*, *29*(2-4), 403–431. doi:10.1080/10549810903550837

McKay, B., & Colque, G. (2016). Bolivia's Soy Complex: The Development of 'Productive Exclusion'. *The Journal of Peasant Studies*, *43*(2), 583–610. doi:10.1080/03066150.2015.1053875

McKay, B.M. (2020). Food Sovereignty and Neo-Extractivism: Limits and Possibilities of an Alternative Development Model. *Globalizations*, *17*(8), 1386–1404. doi:10.1080/14747731.2019.1691798

McKay, B.M. (2017). The Politics of Agrarian Change in Bolivia's Soy Complex. *Journal of Agrarian Change*, *18*(2), 406–424. doi:10.1111/joac.12240

McKay, B.M. (2018). The Politics of Convergence in Bolivia: Social Movements and the State. *Third World Quarterly*, *39*(7), 1247–1269. doi:10.1080/01436597.2017.1399056

McNelly, A. (2020). The Incorporation of Social Organizations under the MAS in Bolivia. *Latin American Perspectives*, *47*(4), 76–95. doi:10.1177/0094582X20918556

Miranda Chávez, C. G. (2016). *Análisis sobre la aplicación del modelo de la gestión compartida en las áreas protegidas nacionales de Bolivia: avances, estancamiento y perspectivas* (Unpublished master's dissertation). San José, Costa Rica: Universidad para la Cooperación Internacional.

Müller, R., Pacheco, P., & Montero, J.C. (2014). The Context of Deforestation and Forest Degradation in Bolivia: Drivers, Agents and Institutions. Occasional Paper 80, Center for International Forestry Research (CIFOR). Retrieved from https://www.cifor.org/publications/pdf_files/OccPapers/OP-108.pdf

Müller, R., Pistorius, T., Rohde, S., Gerold, G., & Pacheco, P. (2013). Policy Options to Reduce Deforestation Based on a Systematic Analysis of Drivers and Agents in Lowland Bolivia. *Land Use Policy*, *30*, 895–907. doi:10.1016/j.landusepol.2012.06.019

Murcia, C., Guariguata, M.R., Peralvo, M., Gálmez, V. (2017). La restauración de bosques andinos tropicales: avances, desafíos y perspectivas del futuro. Documentos ocasionales 170, Centro para la Investigación Forestal Internacional (CIFOR). Retrieved from https://www.cifor.org/publications/pdf_files/OccPapers/OP-170.pdf

Notiboliviarural (2019, January 24). Comunarios piden renuncia de Rocha y la autoridad asegura que mejorará control de bosques. Retrieved from https://www.notiboliviarural.com/mas/actualidad/comunarios-piden-renuncia-de-rocha-y-la-autoridad-asegura-que-mejorara-control-de-bosques

Opinión (2019, January 17). Afines al MAS rechazan designación de Cliver Rocha en la ABT. Retrieved https://www.opinion.com.bo/articulo/el-pais/afines-mas-rechazan-designaci-oacute-n-cliver-rocha-abt/20190117101300638597.html

Orellana Halkyer, R. (2019). ¿Política ambiental o desarrollo integral en armonía con la Madre Tierra? Una mirada reflexiva sobre la configuración de las políticas ambientales en Bolivia. In M. Inturias, K. von Stosch, H. Baldelomar, & I. Rodríguez (Eds.), *Bolivia: Desafíos Socioambientales en las tierras bajas* (pp. 79–98). Santa Cruz de la Sierra: Instituto de Investigación Científica Social (IICS) de la Universidad Nur.

Ormachea, S. E., & Ramirez, F. N. (2013). Políticas agrarias del gobierno del MAS o la agenda del "poder empresarial-hacendal". Documento de coyuntura n.° 19. La Paz: Centro de Estudios para el Desarrollo Laboral y Agrario – CEDLA.

Pacheco Balanza, D. (2014). *Una mirada a la política de bosques en Bolivia: para la descolonización de las políticas*. La Paz: Universidad de la Cordillera.

Pacheco, P., & Benatti, J. H. (2015). Tenure Security and Land Appropriation under Changing Environmental Governance in Lowland Bolivia and Pará. *Forests*, *6*, 464–491. doi:10.3390/f6020464

Ranta, E. (2018). *Vivir Bien as an Alternative to Neoliberal Globalization: Can Indigenous Terminologies Decolonize the State?* New York: Routledge.

Redo, D., Millington, A. C., & Hindery, D. (2011). Deforestation Dynamics and Policy Changes in Bolivia's Post-Neoliberal Era. *Land Use Policy*, *28*, 227–241. doi:10.1016/j.landusepol.2010.06.004

Roca, M. (2018, December 24). El director de la ABT deja el cargo por problemas de salud. *El Deber*. Retrieved from https://eldeber.com.bo/economia/el-director-de-la-abt-deja-el-cargo-por-problemas-de-salud_119088

Romero-Muñoz, A., Fernández-Llamazares, A., Moraes R.M., Larrea-Alcázar, D. M., & Wordley, C.F.R. (2019). A Pivotal Year for Bolivian Conservation Policy. *Nature Ecology & Evolution*, *3*, 866–869. doi:10.1038/s41559-019-0893-3.

Salazar Lohman, H. (2020). Revisiting Bolivian "Progressivism": The Anticommunalism of the Plurinational State. *Latin American Perspectives*, *47*(5), 148–162. doi:10.1177/0094582X20933637

Servindi (2019, June 7). Bolivia: destapan corrupción con madera en Autoridad de Bosques. Retrieved from https://www.servindi.org/actualidad-noticias/07/06/2019/destapan-un-millonario-caso-de-corrupcion-con-madera-en-la-abt

Sierra Praeli, Y. (2019a, July 25). Bolivia: polémica norma pone en peligro cuatro millones de hectáreas en la Amazonía. *Mongabay Latam*. Retrieved from https://es.mongabay.com/2019/07/bolivia-polemica-norma-en-la-amazonia/

Sierra Praeli, Y. (2019b, April 9). Polémica en Bolívia: gobierno decide ampliar frontera agrícola en 250 000 hectáreas para soya transgénica.

Mongabay Latam. Retrieved from https://es.mongabay.com/2019/04/bolivia-gobierno-soya-transgenica/

Soliz Tito, L. (2015). *Cumbre agropecuaria "Sembrando Bolivia": resultados, ecos y primeros pasos hacia su implementación*. La Paz: Centro de Investigación y Promoción del Campesinado.

Springerová, P., & Vališková, B. (2016). Territoriality in the Development of Evo Morales' Government and Its Impacts on the Rights of Indigenous People: The Case of TIPNIS. *Canadian Journal of Latin American and Caribbean Studies*, *41*(2), 147–172. doi:10.1080/08263663.2016.1182297

Unitel (2014, June 1). ABT: La deforestación entre 2010 y 2013 se redujo en 75% gracias al aumento de la multa por hectárea deforestada. Retrieved from https://unitel.bo/economia/abt-la-deforestacion-entre-2010-y-2013-se-redujo-en-75-gracias-al-aumento-de-la-multa-por-hectarea-deforestada_81012

Villalobos, G. (2020, February 20). Las leyes incendiarias en Bolivia. *Fundación Solón*. Retrieved from https://fundacionsolon.org/2020/02/20/las-leyes-incendiarias-en-bolivia/#_ftn4

Webber, J. R. (2016). Evo Morales, *Transformismo*, and the Consolidation of Agrarian Capitalism in Bolivia. *Journal of Agrarian Change*, *17*(2), 330–347. doi:10.1111/joac.12209

Wolff, J. (2020). The Turbulent End of an Era in Bolivia: Contested Elections, the Ouster of Evo Morales, and the Beginning of a Transition Towards an Uncertain Future. *Revista de Ciencia Política*, *40*(2), 163–186. doi:10.4067/S0718-090X2020005000105

List of Interviews

1. Interview with Bolivian expert on environmental public policy and former advisor to Evo Morales's government, Google Meet, August 2020.
2. Interview with current Bolivian member of a national ENGO, Google Meet, September 2020.
3. Interview with current Bolivian member of a national ENGO, Google Meet, September 2020.
4. Interview with former senior official of the MMAyA, Google Meet, September 2020.
5. Interview with former official of the MMAyA, WhatsApp, September 2020.
6. Interview with current Bolivian member of an international ENGO, Google Meet, September 2020.
7. Interview with conservation biologist working at an ENGO in Bolivia, Google Meet, September 2020.

8. Interview with current Bolivian member of an international ENGO, Google Meet, October 2020.
9. Interview with Bolivian expert on environmental public policy, Google Meet, October 2020.
10. Interview with current Bolivian member of a national ENGO, Google Meet, October 2020.

5 Colombia: A (Minimalist) Peace Worse than War

Colombia, the world's second most biodiverse country, accounts for 6% of the Amazon. The causes of deforestation in the country, including Amazonian deforestation, are not fully understood, as decades of intense conflict and the presence of different armed groups have made it difficult for researchers to observe and analyze deforestation processes (Castro-Nuñez et al., 2017; Hoffman, García Márquez, & Krueger, 2018). There is also, as already seen in the introduction to this book, uncertainty in available deforestation figures. Studying the governance of the Colombian Amazon with a focus on deforestation is thus a challenging exercise, one that was further complicated by the fact that, during the interviews conducted, we found a generalized lack of knowledge or deep uncertainty regarding Amazonian deforestation trends, that is, whether deforestation increased or declined over different periods. For the majority of the time interval studied in this work, most interviewees revealed difficulties in associating an increase or reduction in deforestation, and potential causes for growing or declining deforestation levels, with governmental cycles. There was, nevertheless, widespread consensus on the drivers of the recent, highly reported rampant increase in Amazonian deforestation, to which we will turn to later in this chapter; in addition, a number of scientific studies analyzing the topic have been published over the past two years, which facilitated our political examination of more recent years. Due to diverging trends in deforestation levels presented by different sources, and uncertainty among the interviewees, for the period 2002–2015 (see Appendix A, Figure A.4) we do not assume either an increase or a decline in deforestation in our analysis, instead focusing on discussing possible reasons for both an increase and a reduction in deforestation. We assume a sharp increase in deforestation levels for the period 2016–2019, in line with all data sources consulted and consensus among the interviewees.

In the Colombian Amazon, most deforestation is concentrated in the foothills, near the Andes. Deforestation has been mainly associated with pasture conversion, illegal logging and mining, illicit crop cultivation and forest fires (Fajardo Montaña, 2013). Infrastructure projects, especially roads opening new colonization fronts (Armenteras et al., 2019b; Dávalos, Sanchez, & Armenteras, 2016); state incentives for cattle ranching and extractive activities (Hoffman et al., 2018); the armed conflict (Castro-Nuñez et al., 2017; Negret et al., 2019); drug trafficking and the coca eradication policy (Rincón-Ruiz, Correa, Léon, & Williams, 2016; Rincón-Ruiz & Kallis, 2013); land speculation (Dávalos, Holmes, Rodríguez, & Armenteras, 2014); unequal land distribution and land grabbing and hoarding in a context marked by lack of clarity in land tenure and high levels of rural poverty (Armenteras et al., 2019a; Castro-Nuñez et al., 2017; Fajardo Montaña, 2016); and the country's land governance system, which, as in other Amazonian countries, incentivizes land clearing by linking the allocation of land titles to the demonstration of occupation and exploitation of the area of interest (Hoffman et al., 2018), have had, as shall be explored in this chapter, a significant impact on deforestation.

Yet, the Colombian Amazon is one of the most well conserved in the basin. This may essentially be explained by two factors. First, the implementation of an ambitious Amazonian conservation policy initiated during the administration of President Virgilio Barco (1986–1990), with the creation of 44 indigenous reservations and five protected areas covering more than 16,0000 km^2, a policy that was strengthened with the approval of the 1991 Colombian Constitution, which established that those areas were imprescriptible, inalienable and inviolable, and recognized the cultural and territorial rights of the country's ethnic minorities (Rodríguez Becerra, 2014). In 2019, indigenous reservations and national natural parks covered 51% and 24%, respectively, of the Amazonian territory (Guio Rodríguez & Rojas Suárez, 2019) – these have helped protect most of the region from extractive industries and curb environmental destruction, in spite of growing threats to their integrity throughout the years. Second, a controversial factor – the armed conflict in the region. The conflict discouraged the establishment of economic activities and development projects; the presence of the Revolutionary Armed Forces of Colombia (FARC)[1] in particular prevented access to large swathes of the forest by the government and companies. In addition, violence in the form of military attacks and massacres by illegal armed groups led to massive population displacement from rural to city areas, thus reducing the pressure on the forest. Moreover, in areas under their control, the FARC rebels imposed strict limits on forest clearing on local populations, as standing forests allowed them to hide themselves and their coca plantations, which were, alongside kidnapping and extorsion, critical sources of revenue for the guerrillas. It should be noted, however, that armed groups have also fueled deforestation. Illegal armed groups have, for example, participated in road construction, coca cultivation, mining extraction and land grabs, forcing people to settle in other areas of the forest, which generated new processes of land clearing; violence between the FARC guerrillas, paramilitary groups, illicit armed groups associated with drug trafficking and the armed forces of the Colombian government has also caused the displacement of entire communities within the region, further aggravating the pressure on primary forests (see, for instance, Castro-Nuñez et al., 2017; Hoffman et al., 2018; Murillo-Sandoval et al., 2020; Sánchez-Cuervo & Aide, 2013). Additionally, because it leads to ungovernability, direct armed confrontation has been associated with increased deforestation – the dominant authority (which, in several cases, was the FARC guerrilla movement) loses its ability to control the territory, and the resulting climate of anarchy favors illegal resource exploitation (Negret et al., 2019). Consequently, as observed by Castro-Nuñez (2017), it would be

> rather misleading to make an umbrella statement that "armed conflict is good for preventing" deforestation in Colombia (…) – full stop. (…) The armed conflict may have contributed to preserving some parts of the forests. For some other parts, it's a totally different story.

In the following pages, we examine the political, economic and social factors with impacts on the Amazon region, particularly on deforestation, throughout the period 2002–2019, thus covering the two administrations of both Álvaro Uribe (2002–2006 and 2006–2010) and Juan Manuel Santos (2010–2014 and 2014–2018) as well as the first 17 months of Iván Duque's administration (2018–2019). The discussion is divided into three distinct periods: 2002–2010, 2010–2015 and 2016–2019. The first period is characterized by setbacks in environmental policy in favor of short-term economic growth and violence and displacements in rural areas in a context in which the country's political agenda was almost entirely dominated by security issues and the combat against the guerrillas was the national government's overriding priority. During the second period, in a more stable national context and amid peace negotiations, environmental protection and land and rural development issues rose up in the country's political agenda, and there were some positive advances. The third period is marked by the emergence of new, massive environmental challenges brought by FARC's demobilization following the signing of the peace accord with the Colombian government in 2016.

5.1 2002–2010: Environmental Setbacks, Violence and Displacements

Álvaro Uribe, a son of an eminent landowner and cattle rancher killed by FARC, rose to Colombia's presidency in 2002 with a political project combining the need to systematically and aggressively fight the insurgents with the defense of the rural status quo, supported by large landowners, cattle ranchers, agribusinesses and some drug traffickers with close ties to paramilitary groups. During Uribe's mandates, the country's environmental institutional capacity and policies were weakened in favor of security and short-term economic concerns, and security and economic policies had significant socioenvironmental impact on the countryside.

As observed by Palacio Castañeda (2010), in this period, the Amazon was significantly affected by the President's democratic security policy (DSP) – an aggressive policy aimed at defeating the country's insurgent and criminal groups, and recovering state control of the national territory through the strengthening of military, security and intelligence forces, and the mobilization of citizen action toward the governmental fight against illegal armed groups, with the creation of a vast network of civilian informants (Feldmann, 2012). Uribe denied the existence of an armed conflict in the country, thus ruling out the possibility of negotiating with the rebels, and associated the FARC guerrillas and drug trafficking with terrorism, in line with his foreign policy alliance with American President George W. Bush. Additionally, the region was also affected by a predatory "investor confidence" policy directed at favoring the interests of elites and industrial and extractive national and international companies in the mining, energy and agro-industrial sectors, and promoting infrastructure projects (e.g., *La Marginal de la Selva* road) – aimed at attracting foreign investment and creating a powerful business community in Colombia (Botero & Rojas, 2018; Estrada, 2017; Potter, 2020).

The DSP and the policy of investor confidence were intrinsically linked – the government's militarization project aimed at increasing state presence in strategic regions, the Amazon included, and implementing an economic project based on the privatization of natural resources (Interviews 4 and 12) – territorial consolidation zones defined under the DSP coincided with areas of interest for agribusiness, mining and fossil fuel companies (Ramírez, 2019, Figure 1). During Uribe's administration, there was an exponential increase in exploitation concessions,[2] including within the Amazon (PNUD, 2012). Significant improvements in security allowed for national steady economic growth and encouraged a boom in foreign investment (Feldmann, 2012). As a result, Uribe had very high levels of popular approval – for example, in 2008, when the government had great military successes against the FARC rebels, weakening the guerrillas and improving all aggregate indicators of violence, the president's approval rate was 78%. Such rates of public approval help explain the timid domestic political opposition that Uribe faced when scandals involving the government broke out (Pachón, 2009), as shall be seen.

Let us start by examining setbacks to the country's environmental institutional capacity and policy. In Colombia, environmental policy has been largely dependent on the will of the head of state, partly because of the relative weakness of domestic pressure groups, both for and against environmental conservation action (Bustos, 2018; Rodríguez Becerra, 2009). As a consequence of Uribe's goal of reducing the state's bureaucracy and fiscal deficit, the Ministry of Environment – whose process of consolidation, even though restricted, had been constant since its creation in 1993 – was merged with the Ministry of Development in the new Ministry of Environment, Housing and Territorial Development, which prioritized housing issues and downplayed environmental affairs (Rodríguez Becerra, 2009). In addition, the budget assigned to the ministry, and to the national environmental system in general, within the nation's general budget, was reduced (Rudas Lleras, 2019, p. 41) – notwithstanding the increase in global public spending registered during Uribe's administration – and the institution lost highly skilled technicians and employees. The focus of the government on the DSP inevitably led to a reduction in the amount of resources available for environmental policy. The capacity of the country's environmental authority to represent the national environmental interest vis-à-vis other ministries and interests was thus severely compromised (in 2006, in alliance with the Ministry of Agriculture, the Ministry of Environment even helped promote a forest law that incentivized deforestation; the law was in force until 2008, when it was declared unconstitutional, as it violated the right of indigenous peoples to prior consultation). Moreover, several environmental or environmentally related units of other public entities were made extinct or weakened (e.g., the environmental policy unit in the National Planning Department and the unit in charge of ethnic minorities policy in the Ministry of Interior). On a regional level, the country's autonomous corporations (CARs), which are in charge of administering the environment and natural resources in their jurisdiction, also lost technicians and employees, and were assigned conflicting duties (e.g., the exercise of functions related to the construction of infrastructure for basic sanitation and potable water supply), which deviated economic resources from environmental protection and conservation activities (Rodríguez Becerra, 2009). Reductions in the public budget assigned to CARs for sustainable development – which manage strategic natural resources in the Amazon and other ecologically relevant regions – were particularly concerning, as those corporations, given the modest development of economic activities in their territories, have limited resources of their own (contrary to what happens in more economically dynamic CARs, which have, for example, rent that comes from the energy sector) and thus depend largely on the nation's general budget (Rudas Lleras, 2008). Consequently, fewer public resources means less state presence in the region and fewer, and less efficient, deforestation control actions (Interviews 2, 6, 7 and 8). It should be noted, though, that, as observed by one interviewee (Interview 8), as a result of military action in the region, the 2000s were a period during which state presence in the Amazon was less weak; however, the potential impact that the presence of the state's security forces in the territory might have had on deforestation levels is the subject of controversy.

The years of Uribe's presidency were marked by high levels of conflict, with strong financial and military support by the United States in the framework of Plan Colombia (initiated during the previous administration, with President Andrés Pastrana) – an initiative to combat drug cartels and insurgent groups in Colombia – and Plan Patriota – the greatest military offensive against the FARC rebels in the country's history. The anti-drug policy employed was based on strengthening the armed forces and applying measures for crop eradication, especially aerial fumigation with glyphosate and, to a lesser extent, forced manual eradication. The discourse of the Colombian government was one that linked aerial fumigation and sustainability, "in the sense that more fumigation means less coca production and hence less deforestation" (Rincón-Ruiz & Kallis, 2013, p. 70). During this period, the total area of coca cultivation in the country decreased significantly – one of the interviewees (Interview 8) associated this fact with a potential decrease in deforestation during Uribe's administration. Nevertheless, coca eradication policy was found to have displaced coca production to more remote, less accessible areas of primary forest and to protected areas, and aerial fumigation was associated with human displacement (Rincón-Ruiz et al., 2016; Rincón-Ruiz & Kallis, 2013), which was found to be a strong predictor of deforestation (Sánchez-Cuervo & Aide, 2013; see also Dávalos et al., 2016). Here, it is important to note that forced land dispossession and associated human displacement reached the highest levels in Colombia's history during the period 2000–2008, particularly between 2000 and 2002, under the administration of Pastrana, and 2005 and 2008, under the administration of Uribe. In the latter's case, this was the result not only of the increase in aerial fumigation, but also of strong stigmatization and criminalization of rural populations as *guerrilleros*, drug traffickers and terrorists, under the premise that they were FARC collaborators, by the state's armed forces and the police. The establishment of military and police stations has been linked to land abandonment, and there is strong evidence of alliances between the Colombian army and paramilitary groups, the latter having carried out in this period what is known as a "counter-agrarian" reform through forced land dispossession (Centro Nacional de Memoria Histórica, 2016) – big cattle ranchers were actively involved in the Colombian armed conflict as leaders of paramilitary groups and as promotors and beneficiaries of violent land dispossession (Gutiérrez-Sanín & Vargas, 2017). Land concentration, which has been historically high, intensified during Uribe's administration, particularly from 2005 onwards. In 2002, the Gini index for land inequality was 0.81; this number rose to 0.85 in 2009 (OECD, 2015). As observed by Armenteras et al. (2019a, p. 1497), "[w]ith most smallholders locked out of ownership, land concentration generates an exorbitant demand for clearing more forests". Although paramilitary land dispossession occurred on a much larger scale in regions other than the Amazon, because FARC was particularly strong there – and would continue to be even after Uribe's offensive against it (Vásquez, 2010) – the region was not immune to the phenomenon (Interviews 3, 4, 5, 10 and 11; Centro Nacional de Memoria Histórica, 2016).

Another important variable to consider here is the relationship between, on the one hand, not only forced land dispossession but also appropriation of wasteland, and, on the other hand, development projects and associated expectations of land appreciation[3] (Interviews 2, 3, 5, 6, 9 and 10; Armenteras et al., 2019b; Dávalos et al., 2014); some interviewees (e.g., Interviews 2, 3, 4 and 5) also highlighted the role of violent drug traffickers, including narco-paramilitary groups, in land appropriation/dispossession in the Amazon, particularly for money laundering purposes (in spite of the aggressive eradication strategy employed, Colombia remained the world's main supplier of cocaine, as technical improvements allowed for a much higher production per unit area). The movement of the FARC rebels to other areas of the Amazon as a response to the military offensive against them, opening new paths and roads and displacing people, was also a factor mentioned by some interviewees (Interviews 3 and 10). It should also be noted that mining concessions in the Amazon, even if they were not formalized, promoted illegal mining in the region and favored groups outside the law (Interview 12).

In fact, the institutional and political environment of this period favored, or even legitimized, violent and illegal action (Interviews 6, 9, 10, 11 and 12; Fernández, 2010), and there were not only forced land dispossessions, but also multiple massacres and assassinations of social and ethnic leaders, human rights defenders, syndicalists and civilians in general[4] (Angarita Cañas, 2012) – which explains why the United States, at a time when the correlation of forces between Republicans and Democrats was changing, and American syndicates were attentively following the situation in Colombia, did not approve the free-trade agreement during Uribe's presidency, despite his attempts. As observed by Angulo (2010), the human rights situation worsened significantly during Uribe's administration. The government's interlinked security and development policies benefited economic powers and regional political elites, while peasants, ethnic groups, social leaders, human rights defendants, syndicates, journalists, political opposition and even justice institutions were turned into enemies of the state (Angulo, 2010; Delgado, Restrepo, & García, 2010; Fernández, 2010). As one interviewee (Interview 11) put it, "the political context defines what is admissible and what is not, and at that time there was a cloak of legitimacy over certain groups and actions". During this period, to name just a few examples, protest was equated to armed subversion by the president (LeGrand, van Isschot, & Riaño-Alcalá, 2017). Indigenous peoples were accused by the government of being Colombia's largest landowners, and advances in indigenous rights were very limited (Interviews 3 and 13; González Posso, 2011); in 2007, the Colombian government even abstained in the voting of the United Nations Declaration on the Rights of Indigenous Peoples. The executive often attacked the judicial branch of government, particularly regarding investigations of "parapolitics", that is, collaboration of congressmen, senators and governors with paramilitaries, accusing the judicial power of acting in favor of the country's terrorists (Angulo,

2010; Pachón, 2009). Uribe refused bills presented by the liberal sectors for state recognition of the victims of violence and land dispossession associated with the armed conflict – including violence by the police and the national army – on the grounds that they constituted an attack on the DSP (Sarmiento, 2010). Violent land dispossession by paramilitary groups[5] was not sanctioned; on the contrary, there were decisions and initiatives oriented toward favoring the legalization of the land of paramilitaries and other semi-legal or illegal actors (García Trujillo, 2021; Gutiérrez-Sanín, 2010).

During Uribe's administration, the historical regressive character and anti-peasant bias of agricultural/agrarian policy in the country was heightened. Land access, agricultural supports and provision of public goods for small farmers decreased substantially, while large landowners and agribusiness benefited. For example, in 2009 the press discovered that *Agro Ingreso Seguro*, which was the government's biggest and most famous program to support the rural poor, was used to finance wealthy farmers, among them being supporters of the president's campaigns in 2002 and 2006; moreover, it was found that several technical barriers prevented the rural poor from accessing the program's subsidies (see García Trujillo, 2021 and Gutiérrez-Sanín, 2010). Additionally, processes of formation and strengthening of Peasant Reserve Areas (ZRC) were suspended – for Uribe, ZRC were zones of terrorism (La República, 2013) – and opportunities for citizen participation in local governance were very limited (Interview 18).

It is interesting to note, though, that Uribe's presidency ranks third in terms of the extension of protected areas created in the Amazon, with more than 22,000 km^2 of protected forest, behind Virgilio Barco (nearly 47,000 km^2) and Juan Manuel Santos (almost 30,000 km^2).[6] Uribe's administrations used the creation of protected areas as a means of institutional appropriation of the Amazonian territory, and linked the region's national parks to the DSP. New protected land in territories of strong FARC presence was declared; furthermore, as the rebels took refuge in national parks, where fumigation was prohibited, the government became vocal regarding deforestation caused by the guerrillas' coca crops in those areas, thus reinforcing the conservationist tone it gave to the state fight against the rebels (Palacio Castañeda, 2010). Yet, that same government did not refrain from fumigating the national park of La Macarena (Samper Pizano, 2006) – having pressurized the country's state council to authorize it (La Silla Vacía, 2014) – in spite of all the environmentally damaging effects associated with glyphosate. Fumigation in protected areas did not achieve greater levels due to pressure by the European Union, which threatened to suspend financing to Colombia's national parks (Palacio Castañeda, 2010) – the country's protected areas system was largely dependent on international cooperation resources, as protected areas were far from being a priority within the nation's general budget (Interviews 2 and 16). Simultaneously, the Uribe administration also linked protected areas to its policy of creating a powerful business community in the country by concessioning natural parks to the private sector (Palacio Castañeda, 2010).

Another way that the DSP was linked to environmental conservation was through the creation of social programs such as the Forest Warden Family program (*Programa Familias Guardabosques*) as part of the government's effort to build a national network of informants against FARC. This program was pointed by an interviewee (Interview 8) as a positive measure that could help explain a potential decrease in deforestation during this period. However, according to Palacio Castañeda (2010), it gave rise to unforeseen processes of deforestation – to access the government's monetary resources, potential beneficiaries had to prove that they were working toward coca eradication, and this led to people slashing and burning areas of the forest to reach those resources.

For the academics studying the complex relationship between the Colombian armed conflict and deforestation that we were able to interview (Interviews 14, 19 and 20) as well as for a Colombian public official from the Ministry of Environment working for decades in the Amazon region (Interview 18), the most robust explanation for a decline in Amazonian deforestation rates during Uribe's administration lies in the very low levels of population density in the region associated with the escalation of violence since the late 1990s – with historical high levels for most of the period addressed in this subsection due to the government's aggressive offensive against the rebels. This translated into massive migration to urban areas. In fact, the Amazonian migratory balance was markedly negative during the mid-1990s and the 2000s (Roca, Bonilla, & Jabba, 2013). In spite of the fact that higher levels of armed conflict have been associated with higher deforestation rates due to, as previously seen, lack of governance – and Uribe's years in power were undoubtedly the ones during which FARC had the worst conditions with which to govern the region (Interviews 3 and 5) – violence was so intense that it caused mass migration from the Amazon. The effect of this migration in reducing pressure on the forest might have been critical for limiting forest clearing and compensating for factors negatively affecting the region at a time when several deforestation-promoting conditions converged (Interview 19); additionally, the very high intensity conflict and the expansion of FARC into new areas of the Amazon prevented many extractive companies from investing in the region (Interviews 7 and 19).

In sum, factors that could have led to an increase in deforestation during this period are (a) the weakening of the country's institutional environmental capacity; (b) the implementation of the investor confidence policy and the promotion of development projects resulting in land speculation and illegal mining activities; (c) aerial fumigation, stigmatization of rural populations, a political and institutional environment favoring violent and illegal action, and land expropriation by illegal armed groups translating into human displacement to other areas of the forest and further land concentration; and (d) less control of the Amazonian territory by FARC. Factors that could have led to a decline in deforestation are (a) the creation of protected areas and (b) the high-intensity of the armed conflict leading to mass migration from the Amazon and the discouragement of extractivist investments. Ambivalent factors include (a) the presence of the Colombian security forces in the region, which could have resulted both in the surveillance of deforestation activities and human displacement; (b) the reduction of the total area of coca cultivation which, however, having been accomplished through aerial fumigation, led to cultivation displacement; and (c) the Forest Warden Family Program, which could have helped protect some parts of the forest, but which also led to unforeseen processes of deforestation.

Amid uncertainty, at least one thing seems certain: if Amazonian deforestation declined during Uribe's administration, that decline was clearly not the result of a governmental conservation policy or a progressive rural strategy.

5.2 2010–2015: Progress Amid Contradictions and Opposition to Peace

In 2010, following Uribe's unsuccessful attempt – blocked by the country's constitutional court – to run for a third consecutive term for Colombia's presidency, Juan Manuel Santos, who was Minister of Defense between 2006 and 2009, was elected president with the support and votes of *uribismo*. However, once in post, Santos broke with Uribe's positions essentially by (a) recognizing the existence of an armed conflict in the country and initiating peace negotiations with FARC, as the correlation of forces between the state and the guerrillas had changed significantly; (b) recomposing bilateral relations with Colombia's neighbor countries, as Santos knew that regional support, and that of Venezuelan President Hugo Chávez in particular, was key to helping him with the peace process (Uribe's narrow alignment with the United States and the excesses of the DSP in a context in which left governments were rising to power in the region had led to a degradation of Colombia's regional relations); (c) approving the Victims and Land Restitution Law (1448/2011) and issuing a decree (4633/2011) on territorial rights restitution for indigenous communities; and (d) restoring cordial relations between the executive and the judiciary. Contrary to Uribe, Santos would govern by consensus.

With a much less aggressive, more conciliatory posture, President Santos was able to conclude the free-trade agreement with the United States in 2011, an important victory that helped him create favorable conditions for advancing his democratic prosperity policy (DPP), which was based on what his government called "development locomotives", that is, the mining, energy, agribusiness, infrastructure and innovation sectors (Interview 5) – at economic and development policy levels, Santos kept the extractivist strategy of his predecessor. Nevertheless, the President's stance in relation to FARC and the peace process would earn him fierce opposition from Uribe and his supporters, and weaken his presidency.

Santos's presidency was an ambiguous one on issues pertaining to the environment. The Ministry of Environment regained its autonomy with the creation of the new Ministry of Environment and Sustainable Development, which allowed it to devote greater attention to the country's environmental needs and improve environmental management (Interview 6). In addition, the ministry's budget was increased (Rudas Lleras, 2019, p. 41) and its technical capacity enhanced (Paz Cardona, 2018). However, the budget remained very low compared with that of the first years of the ministry's existence. Moreover, the ministry in general did not recover the level of technical capacity it had before the administration of Uribe took office (a notable exception is the Institute of Hydrology, Meteorology and Environmental Studies (IDEAM), as shall be seen), and frequent changes in senior officials compromised the institution's capacity to consolidate itself and advance the country's environmental agenda as expected (Paz Cardona, 2018). During the eight years of Santos's administrations, there were six Ministers of Environment with very different profiles, most of them with no link to, or knowledge of, the environmental sector – the Ministry of Environment was used as an outlet for political tensions, to give representation in government to certain parties and interests

(Interviews 2 and 6). The new ministry was also weaker than the one created in 1993, mainly because decisions regarding land use planning were assigned to the Ministry of Housing and decisions on environmental licensing were transferred to a specialized agency created in 2011; moreover, in 2014, confronted with the end of the commodity rising price cycle, the government established what became known as the "express licensing system", making it easier for mining and hydrocarbon extraction companies to obtain environmental licenses (Rodríguez Becerra, 2019a) – Colombia is highly dependent on the export of non-renewable commodities, particularly oil and coal. Regarding CARs, the budget assigned to them within the nation's general budget remained very low (Rudas Lleras, 2019, p. 56).

During the Rio+20 environmental conference in 2012, Santos announced the formation of strategic mining reserves in an area more than 17,6000 km^2, mainly in the Amazon and Pacific regions, whose concessions would be given to companies that could develop "environmentally sustainable mining". The initiative was presented by the president as a bold measure of environmental protection, but it obviously attracted deep criticism, and the government ended up by declaring a moratorium on mining in the Amazon a couple of months later, which established a number of conditions that would need to be fulfilled before moving forward (Laborde, 2012). Consequently, there were no major legal mining projects in the Amazon, but illegal mining continued to expand. In 2015, a governmental plan for fighting the problem on a national level was presented (Semana, 2015); nevertheless, its results would be modest due to corruption at local level (Interview 12).

Regarding the creation of forest protected areas, the Santos administration's record is impressive, having more than doubled Colombia's protected territory in eight years (Paz Cardona, 2018). In the Amazon, as we have already seen, approximately 30,000 km^2 of new protected forest areas were created, with the expansion of the Chiribiquete National Park – situated in the heart of the region – in 2013 and again in 2018, after complex negotiations with the Ministry of Mines and Energy, during which the Ministry of Environment had the president's support; indigenous reservations would also be expanded throughout the park's perimeter. This was an important sign by the government which frustrated expectations created by mining and hydrocarbon concessions already granted in the territory (Interviews 6 and 18). Moreover, the country's Forest and Carbon Monitoring System (SMByC) was created within IDEAM in 2012 – the monitoring of Colombia's forests emerged as a priority in both national development plans of the president's two administrations, and both contained strategic lines aimed at protecting and conserving forests and forest ecosystems – and the institute's technical capabilities were strengthened. This led to the production of more robust information to inform policy-making (Interviews 6 and 14; Gómez Lee & Quiroga Cubillos, 2019). There was, however, a worrying lack of coherence between conservation and energy policies – in 2017, according to a report by the Contraloría General de la República (2017), 18 areas of the country's system of protected areas overlapped with 29 oil blocks in execution and an additional 12 overlapped with 15 blocks that were already reserved or available. In the national development plan of Santos's second mandate, and other plans derived from it at regional level, for example, there were a number of mid-term sectorial planning instruments determining long-term tendencies which pointed toward the exploitation of the Amazonian energy, mining and agro-industrial potentials as well as to the expansion of transport infrastructure (Botero & Rojas, 2018). Such incoherencies reflected, on the one hand, Santos's strategy to enhance Colombia's international prestige and win the support of the international community for the peace process, within which the environment and environmental diplomacy were an important dimension, and, on the other hand, Colombia's heavy dependence on commodities and the President's unstable multi-party coalition, as addressed below in this subsection.

As for the armed conflict, violence declined significantly in this period. This allowed for greater (non-military) state and ONG presence in the Amazon, and opened new possibilities for working with local, including indigenous communities; collaboration between the Ministry of Environment and its institutions and civil society organizations on the ground increased (Interviews 1, 2 and 6). In addition, FARC had much greater control over the region in this period compared to the previous one, which might help explain a potential decline in deforestation (Interviews 3, 5 and 10) – after Uribe's offensive, FARC's territorial presence and strength reduced substantially but, as previously mentioned, the rebels continued to be strong in the Amazon region. Moreover, aerial fumigation decreased and was suspended in 2015, following pressure by the Minister of Health – some of the interviewees (e.g., Interviews 7, 10, 11, 16 and 17) associated the reduction in aerial fumigation with a potential decline in deforestation, because, as seen in the previous subsection, fumigation has been linked to both cultivation displacement and human displacement. It should be noted, though, that during peace negotiations, FARC incentivized coca cultivation as a means of enhancing its negotiating power, and the area with coca crops in Colombia has been increasing since 2013 (UNODC, 2017). Nevertheless, at least during this period, coca cultivation seemed to have had little influence on deforestation rates (Dávalos et al., 2016). Another point worth noting is that as the peace process developed, so did the oil incursions in the Amazon (Ciro Rodríguez, 2018) – for many in more privileged sectors, peace is simply a way of allowing access to the region to extractivist companies (Interview 7). It is interesting to note, nonetheless, that several interviewees from different sectors (Interviews 1, 4, 9, 10, 11, 14, 15, 16 and 18), indicated the peace process was an important variable in discouraging deforestation, arguing that openness to citizen participation in the process[7] and the expectations created by the peace agreement that was being formulated – which contains important provisions for local development and justice – translated into a "wait-and-see" attitude at local level, with some activities being interrupted as a result, which favored forest protection.

As for the situation of indigenous communities, within the context of the peace process there were positive advances toward recognition of their rights – a decree (1953/2014) creating a special regime for operationalizing indigenous autonomy in the governance of their territories was issued, and later another decree (632/2018) would formally recognize indigenous authority over more than 18,000 km^2 in the Amazon region, where there had been a legal void for almost three decades. President Santos also supported an ambitious proposal by Colombian civil society groups and the Ministries of Environment and Foreign Affairs to create the world's largest ecological corridor – the Triple A (Andes-Amazon-Atlantic Corridor) initiative – designed not only to protect ecosystems but also local populations and indigenous cultures (Interview 13).[8] It should be noted, nevertheless, that his government's relationship with indigenous communities was not without controversy, with attempts to disregard the results of indigenous prior consultations on extractivist projects (Roa Avendaño, 2018).

Concerning agrarian/rural policy, there was, at least in the first years of Santos's first mandate, a significant shift. In 2010, Juan Camilo Restrepo, a figure coming from the political establishment but who was sensitive to the difficult situation in the Colombian countryside, was appointed Minister of Agriculture and made land restitution, land titling and rural development the ministry's priorities. This shift would facilitate negotiations with FARC, as the Colombian armed conflict was fundamentally related to agrarian inequalities; the reformist path adopted, which became visible during the process of approval of the

Victims and Land Restitution Law, would heavily influence the peace talks and the peace agreement. New bills, institutions and policy instruments directed at advancing the new agenda of the Ministry of Agriculture were promoted; by the end of Santos's presidency, investments in rural development would more than double compared to the previous period. It was never Santos's intent to drastically expropriate land from large landowners or allocate all land to peasants, but from his liberal standpoint, the president saw land property rights and rural development as necessary conditions for economic growth and peace (García Trujillo, 2021). The institutional and political environment of this period was thus less favorable to violent land dispossession, and to human displacement as a consequence, than the previous one (Interviews 10 and 11). The practical results of the reformist policy pursued were, however, modest, as a result of insufficient support within the government and opposition from Uribe and the rural status quo. First, more reformist measures started to create divisions within the government and its parliamentary coalition – one cannot forget that those who supported Santos in 2010 did so believing that he would continue Uribe's policies. Moreover, mainly because of his neoliberal orientation, Santos, although he aspired to do so, was never able to obtain the support of the left or establish a stable relationship with the peasant leaderships either – a relationship that worsened after contradictions between discourse and policy became evident (Gutiérrez-Sanín, 2016). Second, the Minister of Agriculture resigned in 2013 following pressure by his own party, the Conservative Party, which was unhappy with his technocratic-based, rather than party-based, appointments to the ministry. Third, Santos faced strong Uribist opposition to the peace process (García Trujillo, 2021). Following the announcement of peace talks in 2012, Uribe's-led opposition mobilized to delegitimize peace in the eyes of the public, which led to widespread underestimation of Santos's efforts to end the armed conflict (Rodríguez-Raga, 2017). The ex-president stated on numerous occasions that the government was negotiating with terrorists and allowing murderers to get away with their crimes and participate in political decision-making processes; that peace was a threat to private property and would lead to massive land expropriation; that the country's security conditions were deteriorating; or that the agreements were permissive in relation to coca crops and drug trafficking (Uribe Vélez, 2014). To institutionalize his opposition, Uribe created the Democratic Center Party, which was elected as the main opposition party in 2014. Santos was thus facing declining levels of public approval, which were aggravated by massive peasant mobilizations against the opening of the country to global food markets in 2013 and 2014. Consequently, the president became increasingly dependent on his coalition to govern, being forced to give the Ministry of Agriculture and its agencies to members of his multi-party coalition whose priorities differed significantly from those of Restrepo (García Trujillo, 2021). As a result, there were agrarian legislative initiatives allowing big landowners, industrialists and financiers to buy wasteland; revoking the limits on purchasing lands in certain regions, which in some cases benefited well-connected politicians (Gutiérrez-Sanín, 2016); and creating special agro-industrial zones (ZIDRES), including within the Amazon (Fajardo Montaña, 2016). In a nutshell, to advance the peace agenda, Santos was simultaneously trying to respond to demands from below and to the vested interests of members of his coalition, whose support he needed to put an end to the armed conflict (Gutiérrez-Sanín, 2016). Supporting peace (particularly in light of the business opportunities that it could bring) was one thing; backing an ambitious agrarian transformation was a completely different one – while many did not oppose the former, most opposed the latter.

Despite all obstacles, the peace talks continued, and an agreement would be signed in 2016. International support for peace favored the process on a domestic level. Colombia's foreign policy for this period – flexible, open and diversified – was built precisely around Santos's intention to secure the support of the international community and attract resources for peace. While Uribe's foreign policy strategy was characterized by the internationalization of the armed conflict, Santos's was marked by the internationalization of peace. During his administrations, Colombia participated more actively in international and regional organizations, built new alliances with different actors, embraced new themes and agendas, engaged in dialogue with non-governmental organizations, and presented its candidacy to the Organization for Economic Co-operation and Development (OECD), which would be accepted in 2018. Interestingly, as part of his strategy to raise the country's international profile, Santos sought to build an image of Colombia as an environmental power (see David, Badillo, & Rodríguez, 2019; Sánchez & Campos, 2019). In 2015, in the context of negotiations for the Paris Climate Agreement, the president launched Sustainable Colombia (*Colombia Sostenible*), an initiative designed to attract foreign funding for projects to mitigate deforestation emissions and promote development in the municipalities most affected by the armed conflict (Bustos, 2018), and signed a 100-million-dollar agreement with Norway, Germany and the United Kingdom to reduce deforestation in the Amazon and empower local populations – this would lead in 2016 to the creation of the Amazon Vision program; however, results would be modest, and deforestation would, for the reasons explored in the next subsection, increase exponentially.

In sum, factors that could have led to an increase in deforestation during this period are (a) the deepening of extractivism, the establishment of the express licensing system and incoherencies between conservation and energy and mining policies incentivizing illegal mining and land speculation; (b) the high rotation of Ministers of Environment hindering the state's capacity to implement a solid, coherent conservation policy; and (c) the increase in coca cultivation area. Factors that could have led to a decline in deforestation are (a) the re-granting of autonomy and allocation of more resources to the environmental sector; (b) the strengthening of forest monitoring and the consequent production of more robust data to inform conservation policies; c) the expansion of the Chiribiquete National Park; (d) greater non-military state and ONG presence in the Amazon, and collaboration with local communities; (e) reduction in and later suspension of aerial fumigation; (f) FARC's greater control of the region; (g) more support for rural populations; and (h) an institutional and political environment that was less favorable for land dispossession and displacement. The peace process emerged as an ambivalent factor leading both to an increase in extractivist incursions into the Amazon and a "wait-and-see" approach by local populations, resulting in less deforestation activities.

5.3 2016–2019: Massive Environmental Challenges in a (Minimalist) Peace Context

This period was marked by deep polarization between those opposing the peace agreement or advocating for a minimalist version thereof, and those supporting it. In August 2016, the government and FARC announced that a final agreement had been reached, and President Santos, optimistic about the result, informed the country that a referendum to ratify the deal would be held in October. Uribe's party led the "no campaign" against ratification supported by conservative figures and religious groups for whom the deal with FARC contained a dangerous "gender ideology" compromising family values. Their strategy consisted of spreading a number of inaccuracies and lies about the agreement, which was eventually refused with 50.2% of the vote. This was a devastating result for the government. However, following massive demonstrations in favor of peace, the government and FARC worked to reach a new agreement (Rodríguez-Raga, 2017).

Six critical points form the final agreement: (1) a comprehensive rural reform aimed at the structural transformation of the countryside, including mechanisms for land access and formalization, and creating conditions for the well-being of rural populations; (2) plural political participation in the construction and consolidation of peace; (3) the end of the conflict, with the ceasefire, the laying down of arms and the demobilization and reincorporation of the FARC insurgents into civil life; (4) a differentiated solution to the drug problem based on voluntary eradication and incentives for transitioning to legal activities, in harmony with human rights and public health considerations; (5) compensation for the victims of the armed conflict; and (6) mechanisms for the implementation of the agreement and verification of compliance with its provisions. For the reasons addressed in the previous subsections, points 1 and 4 are particularly relevant for forest protection. Although it is generally not understood this way in Colombia (Interviews 2, 4 and 7), the agreement contains a strong environmental dimension. Nevertheless, implementation of most of the points has been minimal thus far (Fajardo-Heyward, 2018; Gutiérrez-Sanín, Machuca Pérez, & Cristancho, 2019), and the only point that was fully implemented, that is, point 3, has had a tremendous negative impact on the Amazonian forest.

The agreement was ratified by Congress in November 2016 – another popular vote would have been very risky, but without it the final deal was accused by many of lacking legitimacy, which deepened divisions, strengthened opposition and made the legislative process to approve the necessary norms to implement the agreement very difficult (Rodríguez-Raga, 2017). The position of Santos and the process of implementation of peace were further weakened by (a) the slowing of economic growth; (b) the fact that the Odebrecht scandal occurred in the country and implicated several Colombian politicians, including the president; (c) prospects and political considerations in light of the 2018 elections leading to the resignations of various ministers, and to members of the government coalition starting to distance themselves from the president; (d) the election of Donald Trump in the United States, with the new American president being less sympathetic than his predecessor to the Santos adminstration's approach to the drug problem, threatening to consider Colombia a non-cooperative country in the fight against narcotics as a result of the increase in coca crops; and (e) the Venezuelan crisis, which generated the greatest migration wave in Colombia's history and gave rise to challenging socioeconomic pressures (Fajardo-Heyward, 2018). In addition, as observed by Rettberg (2020b, p. 95), beyond opposition and the unfavorable context surrounding the peace agreement, "some of (...)[its] aspirations

(…) simply exceed the capacities and inertia of existing national state institutions".

In the Amazon, since the departure of the guerrillas, deforestation has been increasing exponentially. FARC dissidents, criminal gangs, drug-trafficking organizations, guerrillas of the National Liberation Army (ELN), cattle ranchers, land mafias and new settlers invaded the areas previously occupied by the guerrillas, where there are now disputes over wealth and territory and a climate of anarchy favoring deforestation. Investors in search of new land have also started to penetrate the territory. All these actors have been moving in a context marked by corruption, poverty, the promotion of road infrastructure in the framework of the development provisions included in the peace agreement as well as projects and plans for the mining, energy and agro-industrial sectors, and lack of clarity in land tenure, a critical aspect whose resolution is part of point one of the peace agreement, which is precisely that which has advanced the least in terms of implementation. Consequently, land hoarding and cattle ranching have grown exponentially, with criminal groups and landowners hiring the poor to deforest large areas (see, for instance, Arenas, 2018; Armenteras, González, & Barreto, 2018; Botero & Rojas, 2018). In the face of rampant deforestation, a group of children and young people presented a plea to the Colombian Supreme Court, which eventually declared the Amazon an entity subject to rights and ordered the government to act to curb deforestation. However, Santos's efforts – for example, the second expansion of the Chiribiquete National Park, the promotion of the Triple A ecological corridor, the suspension of construction of the *Marginal de la Selva* road, the delimitation of the agricultural frontier, the implementation of the Amazon Vision program or the promotion of community forest management units (Botero & Rojas, 2018) – were insufficient and therefore unsuccessful.

In 2018, the candidate for Uribe's Democratic Center Party, Iván Duque, won the presidential election with 54% of the popular vote against the leftist, pro-peace candidate Gustavo Petro, with a political agenda based on the implementation of a "minimalist peace". This consists, among other things, of largely restricting (a) the political and economic reforms agreed in terms of integral rural development to ensure that agribusinesses encounter no obstacles, and (b) the voluntary substitution of illicit crops with legal activities to resume aerial fumigation in close alignment with the US's anti-narcotics vision and policy (Saffon Sanín & Güiza Gómez, 2019). This has major implications for the Amazon.

It is now clear that regional rural elites seek the installation of postconflict Colombia as an agricultural powerhouse, and that current Amazonian chaos is benefiting powerful economic interests; there are currently convenient alliances in the region between illegal and legal actors (Interviews 1, 3, 4, 5, 9, 10, 13 and 16). As explained by one of the interviewees (Interview 3),

> agro-industrial commodities are the most important factor influencing deforestation right now. Through both illegal and legal means, the necessary conditions for incorporating new land into a new cluster that will include large infrastructure developments and large agro-industrial areas, alongside greater mining and energy projects, are being created. Coca crops, small-scale deforestation, inefficient livestock, etc. are simply necessary stages toward achieving that goal.

In fact, "the magnitude of deforestation is not compatible with the development of small scale agriculture (including coca growing), but rather with extensive economic activities" (Prem, Saavedra, & Vargas, 2020, p. 10).

Violence, which was already on the rise given the proliferation of illegal armed groups has, nevertheless, increased substantially since Duque rose to power, particularly forced displacements and assassinations of social leaders and human rights defenders. Unsurprisingly, violence targets particularly those who promote the transformative peace agenda and is mainly linked to economic interests seeking to exploit the land and natural resources. Against this background, Duque has condemned and criminalized social mobilizations – these, according to the new administration, are promoted and financed by illegal armed groups. Simultaneously, indigenous and peasant communities are again being openly stigmatized – Duque's government made clear its stance regarding the latter when, in December 2018, it abstained from voting on the United Nations Declaration on the Rights of Peasants. Moreover, a political-institutional narrative that denies the armed conflict is once more emerging (Interviews 4, 10, 12; see King & Wherry, 2020; López, 2019; Saffon Sanín & Güiza Gómez, 2019).

Accordingly, it should come as no surprise that in 2019, governmental action to control deforestation consisted of the launching of a military operation (*Operación Artemisa*)

> focused on PAs [protected areas], but not in the agricultural frontier limits in which large deforested patches are mostly occurring.

> Artemisa criminalizes small farmers within PAs rather than addressing the fact that larger deforested patches are outside of PAs and that behind these forest clearings are often outsider investors and non-state actors (Murillo-Sandoval et al., 2020, p. 6).

For the new administration, "tourism is Colombia's new oil" and protected areas are a lucrative part of this new source of national revenue (El Tiempo, 2019).

With a national development plan that deepens the country's extractivist economy and presents nature as a strategic national asset to be protected by the military and the police, within the framework of its defense and security policy for legality, entrepreneurship and equity (DSPLEE), the administration of Duque has securitized conservation issues and militarized environmental management, with the mid- to long-term objective of replacing illegal extractivist activities by legal ones granted by the state (Gudynas, 2019). In the meantime, attacks on those who form the most vulnerable part of the deforestation chain are being legitimized in the name of environmental conservation, while the powerful actors and interests that lie behind remain untouched and the structural causes of deforestation, that is, poverty, inequality, lack of opportunities, the presence of armed groups and weak state presence, continue unaddressed (FIP, 2020).

Regarding the coca issue, governmental alignment with the United States and the intent to resume aerial fumigation can essentially be explained by the following reasons: (a) the conservative character of Duque's administration and its willingness to demarcate itself from Santos's administration; (b) the existence of a close relationship between drug trafficking and the recent proliferation of organized armed groups in the Colombian territory; (c) the US's growing pressure on Colombia to curb coca production; and (d) the deterioration of Colombian–Venezuelan relations (see Rettberg, 2020a), which is increasingly seen as a threat to Colombia's national security – the government seeks to ensure the support of the United States in case of military confrontation (Interviews 2 and 9; Márquez Restrepo, 2018). As already seen, aerial fumigation causes human displacement, which causes deforestation; therefore, the possibility of returning to the old approach to the drug problem is bad news for the Amazon.

Duque's Minister of Environment during this period, Ricardo Lozano, although having great knowledge of, and experience in, the environmental sector, having been IDEAM's director, accommodated the predatory vision of other sectors of the government and was too timid, sometimes even invisible, in defense of the country's environmental interests. It should be noted, however, that in December 2019, at COP25, the Colombian government signed a joint declaration with Germany, Norway and the United Kingdom committing to reducing annual deforestation to at least 1,550 km^2 until 2022 and to at least 1,000 km^2 until 2025, when the current national development plan points to maintaining national annual deforestation at (the very high) 2017 levels, that is, at approximately 2,200 km^2 (Torrado, 2019). In the same month, the government also increased by 35% the budget for the environmental sector – by far the largest increase of the past 25 years (Rodríguez Becerra, 2019b). Nevertheless, considering that in the first six months of 2020, primary forest loss in the Colombian Amazon reached a level not too far from what was registered for the whole of 2019 (see MAAP, 2020), it is difficult to take the government's commitment to reducing deforestation seriously. Besides, any commitment to curb deforestation cannot be fulfilled through a purely military and discriminatory approach; the protection of the forest must not be achieved at the expense of the rights and well-being of local populations.

Finally, it is interesting to observe that Duque's public disapproval rate in 2019 was very high (69%) and that a wave of protests triggered by frustration about the slow implementation of the peace agreement, the quality of public education and health, corruption, rising levels of violence and also environmental degradation swept Colombia (Rettberg, 2020a). However, the difficult socioeconomic situation brought about by COVID-19 is now absorbing public attention, and the pressure on the forest will likely increase as a result of the economic crisis associated with the pandemic.

Notes

1 A communist guerrilla group created in the 1960s, inspired by the Cuban Revolution, which degraded in alliance with drug trafficking following the collapse of the Soviet Union.
2 Including mining concessions within national parks, an illegal initiative undertaken by Ingeominas, a governmental agency (Pulido, 2011).
3 For example, in the context of the free-trade agreement with the United States, a long-term competitiveness strategy for the Amazon-Orinoquia region was designed which not only contradicted in many aspects environmental conservation imperatives, but also created expectations that led to land appropriation and violent land dispossession (Interview 10).
4 Thousands of Colombians were assassinated by state agents and reported as *guerrilla* casualties to raise numbers and get rewards such as vacation days or points for career promotion – this is known as the "false positives" scandal.
5 Uribe offered a peace deal to the United Self-Defence Forces of Colombia

(*Autodefensas Unidas de Colombia*, AUC) – the main paramilitary group in the country – and in theory, the organization demobilized during his administration. From the ashes of AUC, Bacrim (criminal bands or *bandas criminales* in Spanish) were born. The government recognized the existence of such groups, but minimized their importance and, most importantly, the link between the old paramilitary groups and Bacrim.

6 Information provided by the National Natural Parks of Colombia – Ministry of Environment.

7 The Colombian peace process has developed impressive mechanisms for civil society participation – both the government and FARC recognized that exclusion from political processes and rural poverty were the core drivers of the conflict, and civil society organizations, which had long been advocating for their right to be part of the decision-making process, lobbied for an inclusive peace process (see Herbolzheimer, 2019).

8 The project has been locked since Bolsonaro rose to Brazil's presidency.

References

Angarita Cañas, P.E. (2012). La seguridad democrática: punta del iceberg del régimen político y económico colombiano. In A. Vargas Velásquez (Ed.), *El prisma de las seguridades en América Latina: escenarios regionales y locales* (pp. 25–50). Ciudad de Buenos Aires: CLACSO.

Angulo S.J.A. (2010). Derechos, uribato, bicentenario. *Cien Días*, *70*, 3–5.

Arenas, N. (2018, September 10). Land Hoarding: What Colombia's New Administration Has Inherited. *Mongabay Latam*. Retrieved from https://news.mongabay.com/2018/09/land-hoarding-what-colombias-new-administration-has-inherited/

Armenteras, D., González, T.M., & Barreto, S. (2018). Fuegos y Áreas Protegidas de la Amazonia Colombiana: Cambio en los Motores de Deforestación. *Revista Colombia Amazónica*, *11*, 73–84.

Armenteras, D., Negret, P., Melgarejo, L.F., Lakes, T.M., Londoño, M.C., GarcíaJ., ... & Davalos, L.M. (2019a). Curb Land Grabbing to Save the Amazon. *Nature Ecology & Evolution*, *3*, 1497. doi:10.1038/s41559-019-1020-1

Armenteras, D., Murcia, U., González, T.M., Barón, O.J., & Arias, J.E. (2019b). Scenarios of Land Use and Land Cover Change for NW Amazonia: Impact on Forest Intactness. *Global Ecology and Conservation*, *17*, e00567. doi:10.1016/j.gecco.2019.e00567

Botero, R., & Rojas, A. (2018). Transformación de la Amazonia: Repercusiones del efecto sinérgico entre políticas erráticas e ingobernabilidad. *Revista Colombia Amazónica*, *11*, 9–32.

Bustos, M.C. (2018). What Shape's Colombia's Foreign Position on Climate Change? *Colombia Internacional*, *94*, 27–51. doi:10.7440/colombiaint94.2018.02

Castro-Nuñez, A. (2017, December 20). Armed Conflict Was Not Always 'Good' For Preventing Deforestation in Colombia (Commentary). *Mongabay Latam*. Retrieved from https://news.mongabay.com/2017/12/

armed-conflict-was-not-always-good-for-preventing-deforestation-in-colombia-commentary/

Castro-Nuñez, A., Mertz, O., Buritica, A., Sosa, C.C., & Lee, S.T. (2017). Land Related Grievances Shape Tropical Forest-Cover in Areas Affected by Armed-Conflict. *Applied Geography*, *85*, 39–50. doi:10.1016/j.apgeog.2017.05.007

Centro Nacional de Memoria Histórica (2016). *Tierras y Conflictos Rurales: Historia, Políticas Agrarias y Protagonistas*. Bogotá: CNMH.

Ciro Rodríguez, E. (2018). "Ni guerra que nos mate, ni paz que nos oprima": incursión petrolera y defensa del agua durante las negociaciones y la firma de la paz en el sur de Colombia. *Colombia Internacional*, *93*, 147–178. doi:10.7440/colombiaint93.2018.06

Contraloría General de la República (2017). *Informe sobre el estado de los recursos naturales y del ambiente 2016–2017*. Bogotá: Contraloría General de la República.

Dávalos, L.M., Holmes, J.S., Rodríguez, N., & Armenteras, D. (2014). Demand for Beef is Unrelated to Pasture Expansion in Northwestern Amazonia. *Biological Conservation*, *170*, 64–73. doi:10.1016/j.biocon.2013.12.018

Dávalos, L.M., Sanchez, K.M., & Armenteras, D. (2016). Deforestation and Coca Cultivation Rooted in Twentieth-Century Development Projects. *BioScience*, *66*(11), 974–982. doi:10.1093/biosci/biw118

David, H.G., Badillo, R., & Rodríguez, M. (2019). Evolución de la política exterior de Colombia en el período 2002–2018. *OASIS*, *29*, 57–79. doi:10.18601/16577558.n29.04

Delgado, A., Restrepo, A.M., & García, M.C. (2010). "Que se mueran los feos". *Cien Días*, *70*, 27–30.

El Tiempo (2019, May 27). Uribe critica a Parques Naturales por límites en áreas protegidas. Retrieved from https://www.eltiempo.com/vida/medio-ambiente/alvaro-uribe-arremete-contra-parques-naturales-por-restricciones-en-areas-protegidas-367382

Estrada, V. (2017). Retroceso del régimen de licencias ambientales en Colombia. *Revista Latinoamericana de Derecho y Políticas Ambientales*, *5*, 69–80.

Fajardo-Heyward, P. (2018). Colombia 2017: entre la implementación y la incertitumbre. *Revista de Ciencia Política*, *38*(2), 233–258. doi:10.4067/s0718-090x2018000200233

Fajardo Montaña, D. (2013). La Amazonia colombiana en la geopolítica regional. *Revista Colombia Amazónica*, *6*, 5–16.

Fajardo Montaña, D. (2016). La frontera agraria en la construcción de la paz. *Revista Colombia Amazónica*, *9*, 35–47.

Feldmann, A.E. (2012). Measuring the Colombian "Success" Story. *Revista de Ciencia Política*, *32*(3), 739–752. doi:10.4067/S0718-090X2012000300014.

Fernández, C. (2010). Las amenazas en el gobierno Uribe. *Cien Días*, *70*, 6–7.

FIP (2020). *Fuerzas militares y la protección del ambiente: roles, riesgos y oportunidades*. Bogotá: Fundación Ideas para la Paz.

García Trujillo, A. (2021). *Peace and Rural Development in Colombia: The*

Window for Distributive Change in Negotiated Transitions. New York: Routledge.

Gómez Lee, M.I., & Quiroga Cubillos, L.C. (2019). Balance de las políticas para mejorar la capacidad de monitorear los bosques en Colombia (2010-2018). In C. Soto I. (Ed.), *Seguimiento y análisis de políticas públicas en Colombia* (pp. 115–131). Bogotá: Universidad Externado de Colombia.

González Posso, C. (2011, August 4). La gran mentira del latifundismo indígena. *Revista Semillas*. Retrieved from https://www.semillas.org.co/es/la-gran-mentira-del-latifundismo-ind

Gudynas, E. (2019, February 9). ¿Se militariza la gestión ambiental y territorial? *El Espectador*. Retrieved from https://blogs.elespectador.com/actualidad/embrollo-del-desarrollo/se-militariza-la-gestion-ambiental-territorial

Guio Rodríguez, C., & Rojas Suárez, A. (2019). *Amazonia colombiana: dinámicas territoriales*. Bogotá: Heinrich Böll Stiftung.

Gutiérrez-Sanín, F. (2016). Agrarian Debates in the Colombian Peace Process: Complex Issues, Unlikely Reformers, Unexpected Enablers. In A. Langer, & G. K. Brown (Eds.), *Building Sustainable Peace: Timing and Sequencing of Post-Conflict Reconstruction and Peacebuilding*. Oxford: Oxford University Press. doi:10.1093/acprof:oso/9780198757276.003.0019

Gutiérrez-Sanín, F. (2010). Land and Property Rights in Colombia – Change and Continuity. *Nordic Journal of Human Rights*, *28*(2), 230–261.

Gutiérrez-Sanín, F., Machuca Pérez, D.X., & Cristancho, S. (2019). ¿Obsolescencia programada? La implementación de la sustitución y sus inconsistencias. *Análisis Político*, *97*, 136–160. doi:10.15446/anpol.v32n97.87197

Gutiérrez-Sanín, F., & Vargas, J. (2017). Agrarian Elite Participation in Colombia's Civil War. *Journal of Agrarian Change*, *17*(4), 739–748. doi:10.1111/joac.12235

Herbolzheimer, K. (2019). Negotiating Inclusive Peace in Colombia. In A. Carl (Ed.), *Negotiating Inclusion in Peace Processes* (pp. 48–55). London: Accord 28, Conciliation Resources.

Hoffman, C., García Márquez, J.R., & Krueger, T. (2018). A Local Perspective on Drivers and Measures to Slow Deforestation in the Andean-Amazonian Foothills of Colombia. *Land Use Policy*, *77*, 379–391. doi:10.1016/j.landusepol.2018.04.043

King, E., & Wherry, S. (2020, April 20). Colombia's Environmental Crisis Accelerates Under Duque. *NACLA*. Retrieved from https://nacla.org/news/2020/04/20/colombia-environmental-crisis-duque

Laborde, R. (2012). Precaución socioambiental: moratoria en la minería de la Amazonía. *Cien Días*, *76*, 49–51.

La República (2013, July 13). Uribe: zonas de reserva campesina son "emporios del terrorismo". Retrieved from https://www.larepublica.co/economia/uribe-zonas-de-reserva-campesina-son-emporios-del-terrorismo-2042753

La Silla Vacía (2014, April 4). Once años después, por fin se prohíbe fumigar los parques nacionales. Retrieved from https://lasillavacia.com/queridodiario/once-anos-despues-por-fin-se-prohibe-fumigar-los-parques-nacionales-47029

LeGrand, C., van Isschot, L., & Riaño-Alcalá, P. (2017). Land, Justice, and Memory: Challenges for Peace in Colombia. *Canadian Journal of Latin American and Caribbean Studies*, *42*(3), 259–276. doi:10.1080/08263663.2017.1378381

López, D. (2019). Menos participación, más mano dura: las políticas de participación ciudadana y seguridad del gobierno Duque. *Cien Días*, 31, 30–34.

MAAP (2020, June 3). MAAP # 120: Deforestation in the Colombian Amazon – 2020. Retrieved from https://maaproject.org/2020/colombian_amaz/

Márquez Restrepo, M.L. (2018). La política exterior de Iván Duque en cien días de gobierno. *Cien Días*, *94*, 60–64.

Murillo-Sandoval, P.J., Van Dexter, K., Van Den Hoek, J., Wrathall, D., & Kennedy, R. (2020). The End of Gunpoint Conservation: Forest Disturbance After the Colombian Peace Agreement. *Environmental Research Letters*, *15*(3), 034033. doi:10.1088/1748-9326/ab6ae3

Negret, P.J., Sonter, L., Watson, J.E.M., Possingham, H.P., Jones, K.R., Suarez, C., ...& Maron, M. (2019). Emerging Evidence that Armed Conflict and Coca Cultivation Influence Deforestation Patterns. *Biological Conservation*, *239*, 108176. doi:10.1016/j.biocon.2019.07.021

OECD (2015). *OECD Review of Agricultural Policies: Colombia 2015*. OECD Publishing.

Pachón, M. (2009). Colombia 2008: Éxitos, Peligros y Desaciertos de la Política de Seguridad Democrática de la Administración Uribe. *Revista de Ciencia Política*, *29*(2), 327–353. doi: 10.4067/S0718-090X2009000200005

Palacio Castañeda, G. (2010). Ecología política y gobernanza en la Amazonia: hacia un balance crítico del régimen de Uribe. In G. Palacio Castañeda (Ed.), *Ecología política de la Amazonia: las profundas y difusas redes de la gobernanza* (pp. 27–60). Bogotá: ILSA, Ecofondo, & Universidad Nacional de Colombia (Sede Amazonia).

Paz Cardona, A.J. (2018, August 6). Colombia: el balance ambiental de Juan Manuel Santos y los enormes retos que le quedan a Iván Duque. *Mongabay Latam*. Retrieved from https://es.mongabay.com/2018/08/balance-ambiental-presidente-juan-manuel-santos-retos-ivan-duque-colombia/

PNUD (2012). Estudio de caso. Minería en territorios indígenas del Guainía en la Orinoquia y la Amazonia colombiana. Retrieved from https://justiciaambientalcolombia.org/wp-content/uploads/2013/03/pnud_estudio-de-caso-_minerc3ada-en-el-guainc3ada__2012-1.pdf

Potter, L. (2020). Colombia's Oil Palm Development in Times of War and 'Peace': Myths, Enablers and the Disparate Realities of Land Control. *Journal of Rural Studies*, *78*, 491–509. doi:10.1016/j.jrurstud.2019.10.035

Prem, M., Saavedra, S., & Vargas, J.F. (2020). End-of-Conflict Deforestation: Evidence from Colombia's Peace. *World Development*, *129*, 104852. doi:10.1016/j.worlddev.2019.104852

Pulido, A. (2011, August 11). La escandalosa adjudicación de títulos mineros en parques naturales. *La Silla Vacía*. Retrieved from https://lasillavacia.com/historia/la-escandalosa-adjudicacion-de-titulos-mineros-en-parques-naturales-26448

Ramírez, M.C. (2019). Militarism on the Colombian Periphery in the Context of Illegality, Counterinsurgency, and the Postconflict. *Current Anthropology*, *60*(19), 134–147. doi:10.1086/699970

Rettberg, A. (2020a). Colombia in 2019: The Paradox of Plenty. *Revista de Ciencia Política*, *40*(2), 235–258. doi:10.4067/S0718-090X202000500010

Rettberg, A. (2020b). Peace-Making Amidst an Unfinished Social Contract: The Case of Colombia. *Journal of Intervention and Statebuilding*, *14*(1), 84–100. doi:10.1080/17502977.2019.1619655

Rincón-Ruiz, A., Correa, H.L., Léon, D.O., & Williams, S. (2016). Coca Cultivation and Crop Eradication in Colombia: The Challenges of Integrating Rural Reality into Effective Anti-Drug Policy. *International Journal of Drug Policy*, *33*, 56–65. doi:10.1016/j.drugpo.2016.06.011

Rincón-Ruiz, A., & Kallis, G. (2013). Caught in the Middle, Colombia's War on Drugs and Its Effects on Forest and People. *Geoforum*, *46*, 60–78. doi:10.1016/j.geoforum.2012.12.009

Roa Avendaño, T. (2018, January 3). Colombia: Extractivist Pax vs. Peace with Social and Environmental Justice. *Upside Down World*. Retrieved http://upsidedownworld.org/archives/colombia/colombia-extractivist-pax-vs-peace-with-social-and-environmental-justice/

Roca, A.M., Bonilla, L., & Jabba, A.S. (2013). Geografía económica de la Amazonia colombiana. Documento de trabajo n.° 193, Banco de la República, Centro de Estudios Económicos Regionales (CEER) – Cartagena.

Rodríguez Becerra, M. (2014). El Estado verde: el caso de la política de conservación de la Amazonia y del Chocó biogeográfico colombianos. In C. Forero Pineda, & L. Díaz Matajira (Eds.), *La gestión de lo público: debates y dilemas* (pp. 135–167). Colombia: Universidad de los Andes.

Rodríguez Becerra, M. (2009). ¿Hacer más verde al Estado colombiano? *Revista de Estudios Sociales*, *32*, 18–33.

Rodríguez Becerra, M. (2019a). *Nuestro planeta, nuestro futuro*. Colombia: Penguin Random House Grupo Editorial.

Rodríguez Becerra, M. (2019b, December 21). Positivas noticias ambientales. *El Tiempo*. Retrieved from https://www.eltiempo.com/opinion/columnistas/manuel-rodriguez-becerra/positivas-noticias-ambientales-columna-de-manuel-rodriguez-becerra-445862

Rodríguez-Raga, J.C. (2017). Colombia: país del año 2016. *Revista de Ciencia Política*, *37*(2), 335–367. doi:10.4067/s0718-090x2017000200335

Rudas Lleras, G. (2008). Financiación del Sistema Nacional Ambiental de Colombia: 1995–2006 y Proyecciones 2007–2010. In M. Rodríguez Becerra (Ed.), *Gobernabilidad, Instituciones y Medio Ambiente en Colombia* (pp. 253–302). Bogotá: Foro Nacional Ambiental.

Rudas Lleras, G. (2019). *Visión panorámica de las fuentes de financiación del*

Sistema Nacional Ambiental – SINA. Bogotá: Fundación Foro Nacional por Colombia.

Saffon Sanín, M.P., & Güiza Gómez, D.I. (2019). Colombia 2018: entre el fracaso de la paz y el inicio de la política programática. *Revista de Ciencia Política*, *39*(2), 217–237. doi:10.4067/S0718-090X2019000200217

Samper Pizano, D. (2006, August 6). Uribe II empieza fumigando parques. *El Tiempo*. Retrieved from https://www.eltiempo.com/archivo/documento/MAM-2131002

Sánchez, F., & Campos, S. (2019). La política exterior de Santos: estrategia y diplomacia por la paz. *OASIS*, *29*, 81–104. doi:10.18601/16577558.n29.05

Sánchez-Cuervo, A.M., & Aide, T.M. (2013). Consequences of the Armed Conflict, Forced Human Displacement, and Land Abandonment on Forest Cover Change in Colombia: A Multi-Scaled Analysis. *Ecosystems*, *16*, 1052–1070. doi:10.1007/s10021-013-9667-y

Sarmiento, F. (2010). Víctimas en el cálculo político. *Cien Días*, *70*, 20–24.

Semana (2015, July 31). Gobierno presenta un plan de lucha contra la minería ilegal. Retrieved from https://www.semana.com/gobierno-presenta-plan-para-combatir-mafias-de-la-mineria-ilegal/436897-3/

Torrado, S. (2019, December 12). Colombia se compromete a una ambiciosa reducción de la deforestación. *El País*. Retrieved from https://elpais.com/sociedad/2019/12/11/actualidad/1576094837_925112.html

UNODC (2017). Colombia: Survey of Territories Affected by Illicit Crops – 2016. Retrieved from https://www.unodc.org/documents/crop-monitoring/Colombia/Colombia_Coca_survey_2016_English_web.pdf

Uribe Vélez, A. (2014, November 17). El error de negociar con las FARC. *El Mundo*. Retrieved from https://www.elmundo.es/internacional/2014/11/17/5468cda5ca47417a758b4577.html

Vásquez, T. (2010). La seguridad democrática de Uribe (2002–2010). *Cien Días*, *70*, 8–11.

List of Interviews

1. Interview with Colombian member of an international ENGO, WhatsApp, May 2020.
2. Interview with former senior official of the Colombian Ministry of Environment, Google Meet, May 2020.
3. Interview with current member of a Colombian ENGO, Google Meet, May 2020.
4. Interview with Colombian anthropologist, Google Meet, May 2020.
5. Interview with Colombian expert on agrarian and rural development issues, WhatsApp, May 2020.
6. Interview with former senior official of the Colombian Ministry of Environment, WhatsApp, June 2020.

7. Interview with Colombian environmental historian, Google Meet, June 2020.
8. Interview with former senior official of the Colombian Ministry of Environment, WhatsApp, June 2020.
9. Interview with Colombian environmental lawyer, WhatsApp, June 2020.
10. Interview with former official of the Colombian Ministry of Agriculture, WhatsApp, June 2020.
11. Interview with Colombian political scientist, WhatsApp, June 2020.
12. Interview with former Amazonian indigenous deputy, WhatsApp, June 2020.
13. Interview with former official of the Colombian Ministry of Interior, WhatsApp, June 2020.
14. Interview with Colombian forest engineer, Google Meet, June 2020.
15. Interview with Colombian environmental lawyer, WhatsApp, 18 June 2020.
16. Interview with former official of the National Natural Parks unit of the Colombian Ministry of Environment, June 2020.
17. Interview with geographer specialized in the Colombian Amazon region, Google Meet, June 2020.
18. Interview with current official of the Colombian Ministry of Environment, Google Meet, July 2020.
19. Interview with Colombian conservation biologist, Zoom, Google Meet, July 2020.
20. Interview with climate and land use scientist, Google Meet, July 2020.

6 Conclusion

The Amazon is losing its resilience. As GHG emissions grow and deforestation advances, the forest's vulnerability increases. The "Amazon tipping point" may be nearer than previously thought (Chiaretti, 2021; Staal et al., 2020); crossing it would potentially have devastating consequences. The forest is an enormous carbon reservoir, plays a major role in the hydrological cycle and is the home of unique species of plants and animals. Its rich ecosystems provide critical ecological services that allow human and non-human life to thrive. The Amazon is also home to different traditional communities and hosts extraordinary social and cultural diversity. This exceptional socio-ecological wealth is now threatened by human actions and political decisions that ignore the enmeshment of human social worlds with non-human others.

Nature is not external to politics. As this book has demonstrated, politics and the Earth system are inherently intertwined. Anthropogenic climate change and current Amazonian politics and policies make the possibility of savannization of the forest alarmingly conceivable. Climate change is accelerating, and governance has deteriorated significantly over the past decade in most of the Amazon; the trajectory of deforestation in the Brazilian Amazon over the last 15 years is proof of how much deforestation control depends on good governance. Environmentally destructive activities have intensified in the four countries addressed. In Brazil, a stable political and socio-economic period characterized by the rise of the country's pro-environmental forces and the strengthening of the MMA, with Amazonian deforestation control becoming a political priority for the Lula administration and a public concern, was followed by a turbulent, complex socioeconomic context and political instability, with a new administration focusing on short-term economic growth considerations and a debilitated MMA, which allowed the country's anti-environmentalist forces, particularly the powerful FPA, to reverse progress achieved in the previous years, and illegal activities to proliferate in the Amazon. Amazonian governance deteriorated further over the following years. The FPA consolidated its political power in a context marked by a severe economic recession, the impeachment of Rousseff and the profound weakness of Temer's presidency, whose government was supported by the ruralists at a cost of the issuing of a series of laws and decrees that severely affected the Amazon. Additionally, the MMA's institutional capacity continued to be weakened with profound budget cuts as part of a policy to reduce the country's fiscal deficit, which further affected the supervisory work of pivotal agencies working in the Amazon. Amid high unemployment levels, systemic corruption, the deteriorating state of public health care services and the growth of crime and violence, environmental issues were far from being a priority concern for Brazilians. It was against this background that openly anti-environmentalist and science-sceptic President Bolsonaro rose to power. Severe environmental policy setbacks and the anti-environmentalist rhetoric of the new administration (the Minister of Environment included) – supported by the most conservative factions of the ruralist bloc – have encouraged illicit forest conversion activities and violence against local communities in the Amazon. The country's environmental movement, which rose in the 1980s and whose political strengthening in the second half of the 2000s was key for shifting the paradigm of Amazonian governance in the country, has over the past decade exhibited difficulties in counteracting the force of the agribusiness lobby. Such difficulties are certainly partly related to the complicated political and socioeconomic context in Brazil, but they are probably also associated with the movement's alliance with the PT, namely with the way it affected the movement's identity and strategies, and the 2012 rupture with the party, whose impact has not been scrutinized yet (Pereira & Viola, 2019). In the foreign policy field, even in its most politically progressive periods, the country has fallen short in the necessary action to promote ambitious responses to the climate and biodiversity crises within the UNFCCC and the CBD's negotiations.

In Peru, planning and decision-making processes have since the 1990s been dominated by the super-powerful, conservative MEF, in close alliance with big businesses, particularly in the mining and energy sectors – actors who have pressured for flexibility in land use policy and environmental norms. From the beginning of decentralization in the forestry, land and mining sectors in the late 2000s, regional governments were never given the necessary resources and capabilities for performing their monitoring and executive responsibilities, and no mechanisms for ensuring transparency and effective oversight were developed. A solid, well-conducted process of decentralization was not in the interest of the country's highly influential economic lobbies, which benefit from the state of disorder in the Amazon – a state that allows for the articulation of different actors and activities (informal, illegal and formal) in the region and encourages illegality, which is further facilitated by the country's profound institutional fragmentation and lack of clarity in rules and procedures. A debilitated civil society and a frail, fragmented political opposition, with an individualized Congress and inexperienced politicians, alongside the growing influence of regional parties since the mid-2000s and of *Fuerza Popular* in more recent years, and widespread fear of change leading the country back to the very difficult situation of the past, have further cemented the strength of economic elites and extractivist industries. Additionally, a policy of facilitating land access to large investments has intensified land concentration in the country, which in turn has translated into migration processes to the Amazon and displacements within the region, exacerbating the pressure on the forest. Population retention programs to control migration from the highlands to the Amazon and policies to support family farming have never been on MEF's agenda; resources have been concentrated on promoting agricultural intensification and commercialization, and not on providing farmers with the technical and financial means that would allow them to make use of the land that has already been deforested and develop sustainable agricultural practices. As a result, the agriculture frontier continued to expand. Positive advances came mostly from action by committed Ministers of Environment, who managed to find opportunities to put the environmental agenda on the table. Nevertheless, MINAM has been very limited in its capacity to influence decisions driving forest clearing in the country, and was eventually weakened in the final years of Humala's government when commodity prices were falling, and during Kuczynski's presidency, whose government openly represented the interests of economic elites. Moreover, since the election of Kuczynski, lawlessness in the Amazon has been further favored by a climate of political instability, in addition to corruption scandals and high levels of crime and public insecurity, which exacerbated political and public neglect of environmental issues. Although President Vizcarra's initiatives to fight corruption and crime, including within the Amazon, were important, they were also insufficient.

In Bolivia, the Morales administration's policy of granting land titles in the Amazon to the MAS's supporters from the highlands not only created tensions with local communities, but also exacerbated deforestation. Additionally, despite the political and economic strength of Santa Cruz's elites, who put the brakes on some of the reforms pursued by the government in its early years, the MAS could have had, since Morales's second mandate, some conditions for advancing its campaign promise of profoundly changing the country, but opted instead for allying with the *cruceños* and deepening extractivism. This was the result of both the threat of separatism in the lowlands and the prevalence of the developmentalist faction of the MAS, represented by Vice-President Álvaro García Linera, over the more indigenist and ecologist bloc. García Linera and the elites and middle classes of Santa Cruz shared a key interest, that is, the expansion of resource extraction. The alliance between the government and the *cruceños* consolidated with the fall in global commodity prices, as the former saw in agricultural expansion its new source of revenue for financing the social welfare programs that legitimized the MAS in the eyes of the Bolivian population. Privileged peasants were also incorporated into the alliance. At the same time, small farmers began to rent their lands to agribusiness and rich peasants, performing jobs other than farming in agro-industrial complexes, a situation that aligned their interests with those of agro-industrial groups. In addition, the co-optation of social movements and the government's action to divide and disarticulate indigenous organizations – whose strength in the beginning of the 2000s led Morales to Bolivia's presidency – severely weakened the activity of those groups. The MMAyA was also too incapacitated and absorbed in water governance issues to fight for the Amazon. Against this background, since 2013, a series of laws and decrees incentivizing the expansion of the agricultural frontier and violating the country's protected areas have been issued, and the ABT's director was replaced by another whose vision was aligned with predatory development.

In Colombia, the fight against FARC and the focus on short-term economic growth have diverted political and public attention as well as resources away from environmental policy during Uribe's presidency, resulting in the weakening of the country's institutional environmental capacity. Additionally, the promotion of development projects boosted land speculation and illegal mining activities in the Amazon. Uribe's policy to defeat the rebels also affected the region socially and ecologically as a result of aerial fumigation, the stigmatization of rural populations and a permissive attitude towards the activities of paramilitary groups, which translated into human displacements within the region and further land concentration. Nevertheless, the high-intensity armed conflict of that period, leading to mass migration from the

Amazon and discouraging extractivist investments, may have helped protect the forest from the pressures it suffered throughout the 2000s. During Santos's presidency, there were some positive advances, such as the reduction and suspension of aerial fumigation, more support for rural populations, fewer land dispossessions and displacements, and the strengthening of forest monitoring. There was also, however, a deepening of extractivism and incoherencies between conservation and energy and mining policies, which reflected the president's strategy to enhance Colombia's international prestige and win the support of the international community for the peace process, on the one hand, and Colombia's heavy dependence on commodities and Santos's unstable multi-party coalition, on the other hand. Additionally, the 2016 peace deal and FARC's departure from the Amazon led to the region's invasion by several actors in a context marked by the absence of the state. The provisions of the peace agreement that could have a substantial impact on curbing deforestation have advanced minimally, partly as a result of opposition by Uribe and his supporters. Uribist President Duque is now supporting the regional rural elites' plan to transform Colombia into an agricultural powerhouse, and condemning and criminalizing social mobilizations. Partly due to the weakness of interest groups, environmental policy in the country depends to a very significant extent on the will of the president. It should also be noted that the Ministry of Environment has been too weak since Uribe's presidency.

In sum, in the four countries progressive, pro-environmental forces have been unable to counteract those pushing for environmentally destructive development – the very powerful agribusiness lobby in Brazil and now the Bolsonaro administration; the influential mining and energy sectors, and the logging industry supported by MEF as well as the *Fuerza Popular* party in Peru; the large landowners and agro-industrial elites of Santa Cruz, privileged peasants and the developmentalist faction of the MAS in Bolivia; the forces of *uribismo* in Colombia; and organized crime in all countries. In Bolivia, progressive agents seemed to have had the necessary strength and conditions to promote important advances in the country's socioenvironmental agenda; however, their co-optation or direct debilitation by the MAS government has hindered their reformist potential. Bolivian social movements need to rebuild themselves and recover their independency. In Brazil, it is possible that the difficulties facing the environmental movement are also partly related to its past close collaboration with the PT. More research on the topic is needed. Reflecting on possible pathways for promoting collaboration between the country's environmental movement and the more reformist branches of the agribusiness sector is also critical (Pereira & Viola, 2019). In Peru and Colombia, countries which experienced recent civil wars and were affected by strong guerrillas, and where social mobilizations have been discouraged and repressed, civil society groups are stigmatized and weak, and more reformist ideologies are still significantly discredited, being frequently associated with terrorism, violence and chaos. Progressive socioenvironmental forces in those two countries seem to have a long road ahead of them.

The following years will be challenging. Latin America and the Caribbean has been the region hardest hit by the COVID-19 crisis in terms of public health and economic development (ECLAC, 2020). In fact, we found widespread pessimism about the future of the Amazon among the interviewees in all countries, who believe that extractivism will deepen in the region. Against this background, we hope that this book can serve as a starting point for a broader discussion on the Amazon, and encourages the development of ambitious responses to the challenges facing the forest. In this process, having forest monitoring systems that allow for a more rigorous identification of primary forest loss by anthropogenic action in all Amazonian countries beyond Brazil would help with conducting policy analyses and designing more robust governance strategies. For the Peruvian case in particular, more field data is needed to fully understand deforestation processes in the country. Focusing on possible pathways for building a shared regional approach to control deforestation within the Amazon Cooperation Treaty Organization (OTCA) is also of the utmost importance (see Pereira & Viola, 2020). Finally, any successful response to protect the Amazon must never lose sight of the human dependence on, and enmeshment in, nature. Otherwise, the tipping point could soon become a reality.

References

Chiaretti, D. (2021, February 5). "Savanização da Amazônia já está ocorrendo", diz Nobre. *Valor Económico*. Retrieved from https://valor.globo.com/brasil/noticia/2021/02/05/savanizacao-da-amazonia-esta-mais-proxima-diz-nobre.ghtml

ECLAC (2020). *Preliminary Overview of the Economies of Latin America and the Caribbean*. Santiago: ECLAC. Retrieved from https://repositorio.cepal.org/bitstream/handle/11362/46504/40/S2000880_en.pdf

Pereira, J.C., & Viola, E. (2019). Catastrophic Climate Risk and Brazilian Amazonian Politics and Policies: A New Research Agenda. *Global Environmental Politics*, *19*(2), 93–103. doi:10.1162/glep_a_00499

Pereira, J.C., & Viola, E. (2020). Close to a Tipping Point? The Amazon and the Challenge of Sustainable Development Under Growing Climate Pressures. *Journal of Latin American Studies*, *52*(3), 467–494. doi:10.1017/S0022216X20000577

Staal, A., Fetzer, I., Wang-Erlandsson, L., Bosmans, J.H.C., Dekker, S.C., van Nes, E.H., … Tuinenburg, O.A. (2020). Hysteresis of Tropical Forests in the 21st Century. *Nature Communications*, *11*(1), 4978. doi:10.1038/s41467-020-18728-7

Appendix A

In this Appendix, we present a compilation of deforestation, primary forest loss and tree cover loss data from various sources for each of the Amazonian countries addressed in this book. The fact that the data from the different sources are shown in a single graphic does not mean that the figures are directly comparable. Each entity uses a different methodology for producing the data, and in some cases the same entity has, over time, changed its methodology. Differences in the geographical limits considered; satellite image datasets, resolution and spectral bands used; approaches to dealing with the existence of clouds over the Amazon region; and how deforestation, forest loss and tree cover loss are accounted for as well as annual measurements that may not follow the civil calendar, are among the main factors that lead to different figures.

As mentioned in Chapter 1, since our research focuses on politics and policies, we are particularly interested in the human causes of (primary) forest loss in the Amazon, that is, deforestation. The Brazilian PRODES program is, to our knowledge, the only source of deforestation data for the region, as their technicians conduct manual image photointerpretation to identify deforested areas, which allows for a more accurate identification of primary forest loss by anthropogenic action. Other entities also use the term *deforestation* to label their data; nevertheless, in their case, it is much more uncertain how the methodologies followed could allow for a separation between changes related to agricultural or other land conversion, and other changes that may not actually be deforestation.

For each of the countries studied, we sought to include one governmental data source,[1] alongside independent data by RAISG and the GFW. For the sake of transparency, we opted to include three different datasets from RAISG, published in 2012, 2016 and 2020, as changes over time in the methodology followed resulted in some significantly different trends compared to data previously published. The

2012 and 2016 RAISG data were provided in five-year intervals, not annually; consequently, these data are shown in the graphics as an annual average. In 2020, RAISG presented new, annual data. As for the GFW, despite the platform's figures not strictly reflecting deforestation, we considered that it was important to include another independent source that also provided data for all Amazonian countries, given the existence of incoherent trends in the cases of Peru, Bolivia and Colombia. To reduce heterogeneity between the different approaches, the GFW data provided here were calculated using the same geographical limits as the 2020 RAISG data. Regarding the Bolivian case, it should be noted that we were unable to find ABT data for some years; for this reason, there are data gaps in the respective graphic. The Monitoring of the Andean Amazon Project (MAAP) also provides data for all Amazonian countries addressed in this book; however, because it uses the same data source as the GFW, MAAP's figures are very similar to those presented by the GFW. Consequently, we did not include MAAP data in our graphics.

Despite the different methodologies followed by the various sources, and the fact that some of the graphics may not be easy to read, we found it useful to present all data in a single graphic for each country, as that allows for the building of a general picture of different trends, while also reflecting the complexity that anyone making a political/policy analysis of Amazonian deforestation has to deal with. Additionally, to our knowledge, no other study has compiled such information.

Brazil

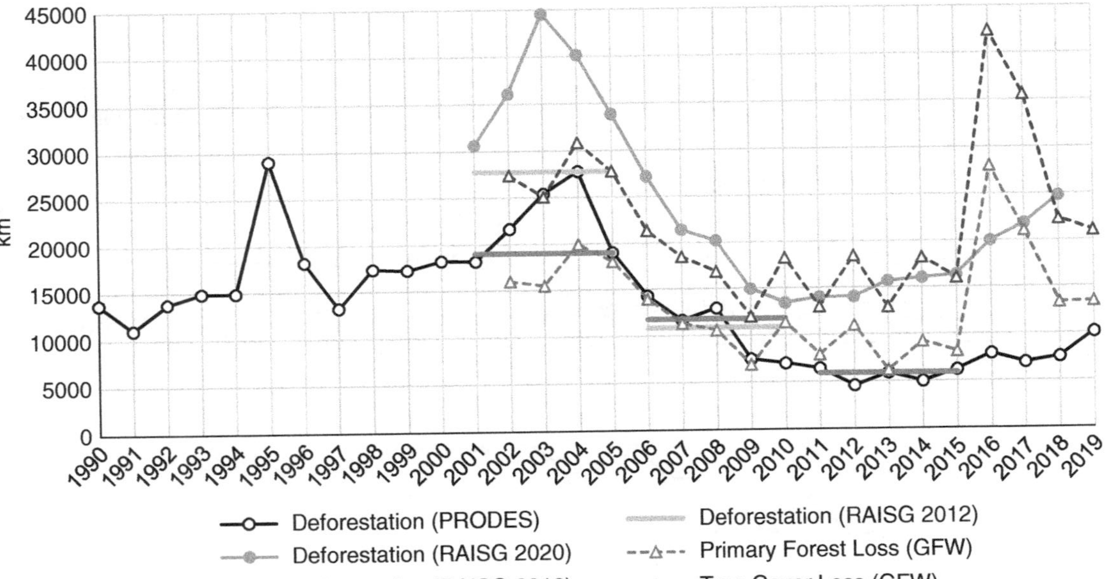

Figure A.1 Deforestation, Primary Forest Loss and Tree Cover Loss Data for the Brazilian Amazon (km^2).
Source: Elaborated by the authors with data from PRODES (2020), RAISG (2012, 2016, 2020) and the GFW.[2]

Peru

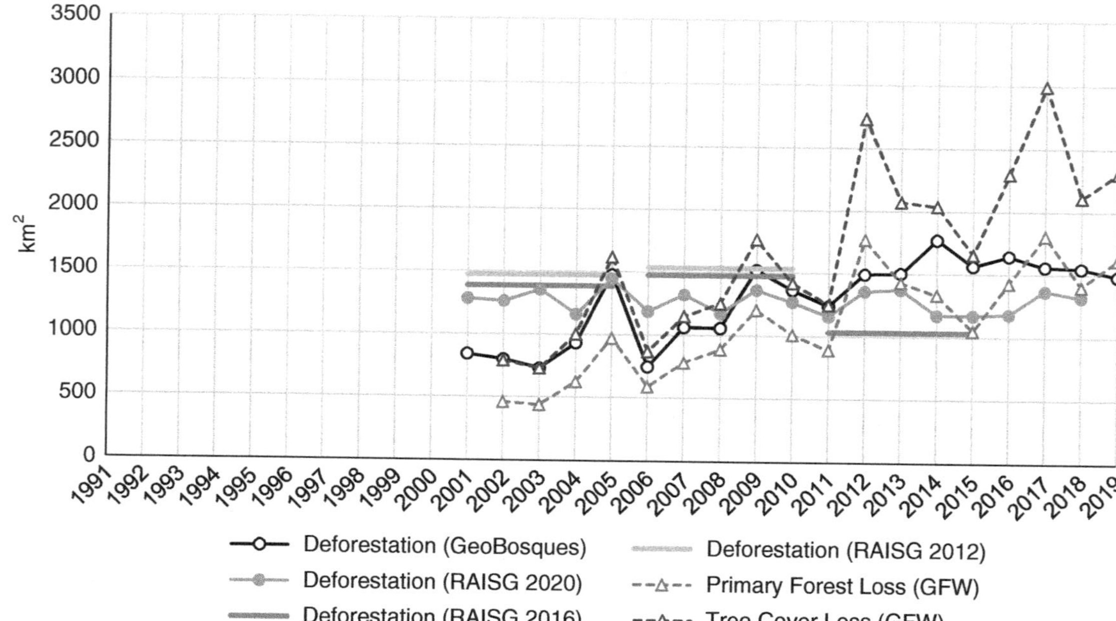

Figure A.2 Deforestation, Primary Forest Loss and Tree Cover Loss Data for the Peruvian Amazon (km^2).
Source: Elaborated by the authors with data from *GeoBosques* (2020), RAISG (2012, 2016, 2020) and the GFW.[3]

Bolivia

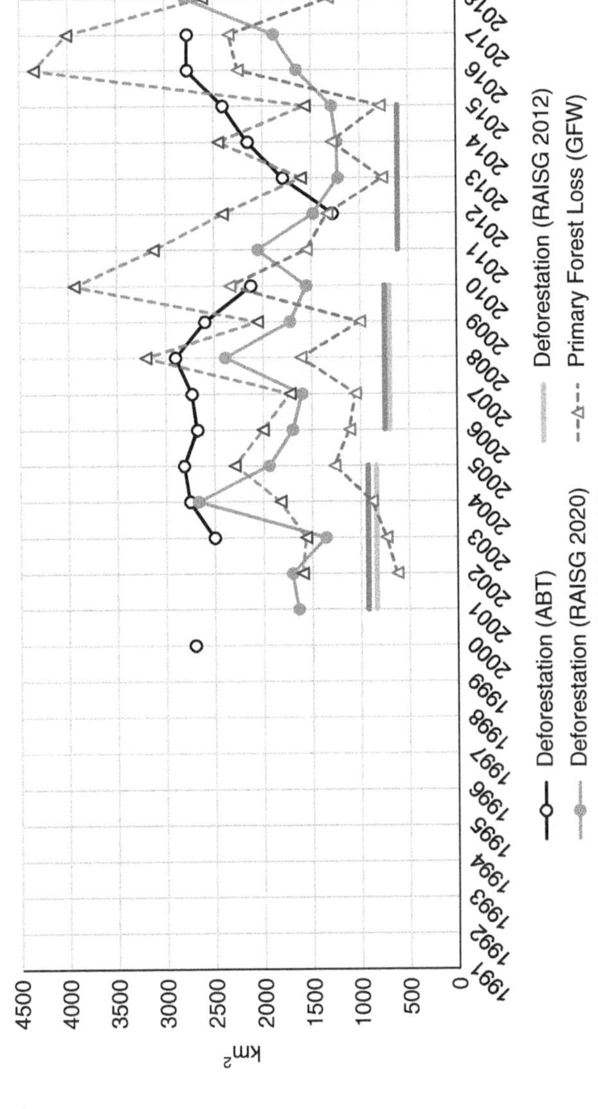

Figure A.3 Deforestation, Primary Forest Loss and Tree Cover Loss Data for the Bolivian Amazon (km²).
Source: Elaborated by the authors with data from ABT (2011, 2016, 2018), RAISG (2012, 2016, 2020) and the GFW.[4]

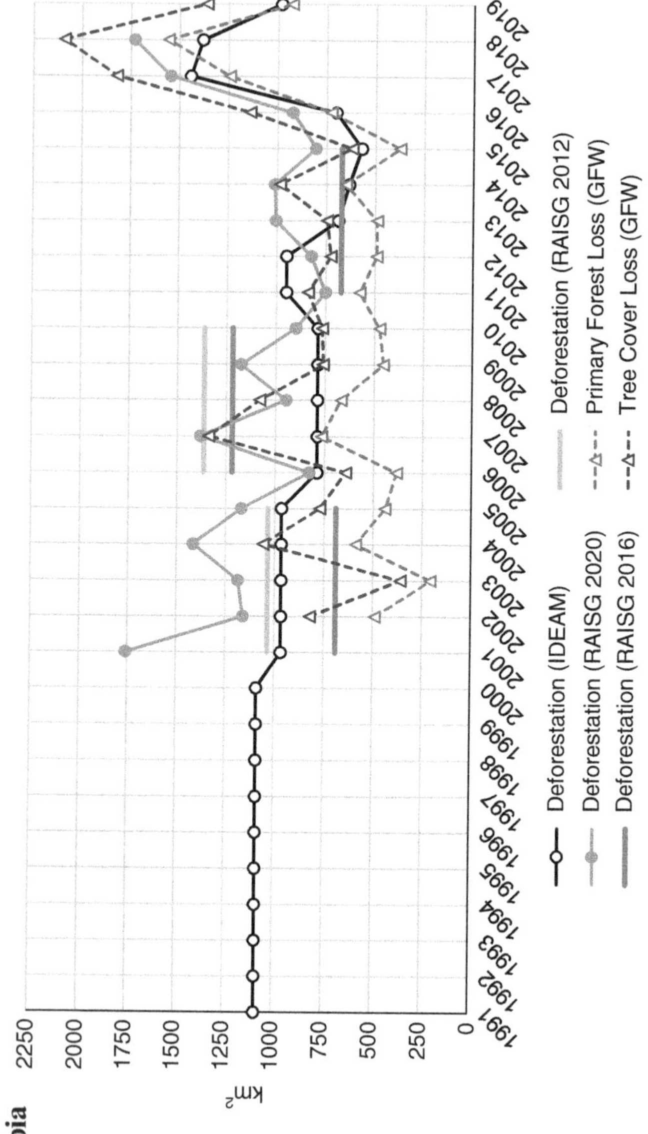

Figure A.4 Deforestation, Primary Forest Loss and Tree Cover Loss Data for the Colombian Amazon (km^2).
Source: Elaborated by the authors with data from IDEAM (2020), RAISG (2012, 2016, 2020) and the GFW.[5]

Notes

1 Although governmental data used for Peru and Bolivia refer to the whole country, there is no substantial difference between figures for the whole country and figures for the Amazon region.
2 The GFW data were provided to us by the platform's staff members, to whom we sent the limits used by RAISG in 2020, which are available here: https://www.amazoniasocioambiental.org/en/maps/#!/download
3 Ibid.
4 Ibid.
5 Ibid.

References

ABT (2016). Deforestación en Bolivia. 2012–2015. Retrieved from http://www.abt.gob.bo/images/stories/Transparencia/InformesAnuales/memorias-2012-2015/Memoria_Deforestaci%C3%B3n_2012_2015_opt.pdf

ABT (2018). Deforestación en el Estado Plurinacional de Bolivia. Periodo 2016–2017. Retrieved from http://www.abt.gob.bo/images/stories/Transparencia/InformesAnuales/memorias-2016-2017/Memoria_Deforestacion_2016_2017_opt.pdf

ABT (2011). *Informe anual 2010 y balance de la década*. Santa Cruz. Retrieved from http://www.abt.gob.bo/images/stories/Transparencia/InformesAnuales/informe_anual_2010_balance_decada.pdf

GeoBosques (2020). Bosque y pérdida de bosque. Retrieved from http://geobosques.minam.gob.pe/geobosque/view/perdida.php

IDEAM (2020). Cambio de la superficie cubierta por bosque natural. Retrieved from http://smbyc.ideam.gov.co/MonitoreoBC-WEB/reg/indexLogOn.jsp

PRODES (2020, November 30). Monitoramento do desmatamento da floresta amazônica brasileira por satélite. Retrieved from http://www.obt.inpe.br/OBT/assuntos/programas/amazonia/prodes

RAISG (2016). *Amazonía 2016 – áreas protegidas y territorios indígenas*. São Paulo: ISA – Instituto Socioambiental.

RAISG (2012). *Amazonia Under Pressure*. São Paulo: ISA – Instituto Socioambiental.

RAISG (2020). Online Map. Retrieved from https://www3.socioambiental.org/geo/RAISGMapaOnline/

Index